Lecture Notes in Artificial Intelligence   11909

Subseries of Lecture Notes in Computer Science

More information about this series at http://www.springer.com/series/1244

Rapeeporn Chamchong ·
Kok Wai Wong (Eds.)

Multi-disciplinary Trends in Artificial Intelligence

13th International Conference, MIWAI 2019
Kuala Lumpur, Malaysia, November 17–19, 2019
Proceedings

 Springer

Editors
Rapeeporn Chamchong ⓘ
Mahasarakham University
Maha Sarakham, Thailand

Kok Wai Wong
Murdoch University
Murdoch, WA, Australia

ISSN 0302-9743   ISSN 1611-3349   (electronic)
Lecture Notes in Artificial Intelligence
ISBN 978-3-030-33708-7   ISBN 978-3-030-33709-4   (eBook)
https://doi.org/10.1007/978-3-030-33709-4

LNCS Sublibrary: SL7 – Artificial Intelligence

This Springer imprint is published by the registered company Springer Nature Switzerland AG
The registered company address is: Gewerbestrasse 11, 6330 Cham, Switzerland

Preface

The Multi-disciplinary International Conference on Artificial Intelligence (MIWAI), formerly called the Multi-disciplinary International Workshop on Artificial Intelligence, is a well-established scientific venue in the field of artificial intelligence (AI). The MIWAI series started in 2007 in Thailand as the Mahasarakham International Workshop on Artificial Intelligence and has been held every year since then. It has emerged as an international workshop with participants from around the world. In 2011, MIWAI 2011 was held outside of Thailand for the first time, in Hyderabad, India, so that it became the "Multi-disciplinary International Workshop on Artificial Intelligence." Then the event took place in various Asian countries: Ho Chi Minh City, Vietnam (2012); Krabi, Thailand (2013); Bangalore, India (2014); Fuzhou, China (2015); Chiang Mai, Thailand (2016); Brunei Darussalam (2017); and Hanoi, Vietnam (2018). In 2018, MIWAI was renamed to the "Multi-disciplinary International Conference on Artificial Intelligence."

The MIWAI series of the conferences serve as a forum for AI researchers and practitioners to discuss and deliberate cutting-edge AI research. It also aims to elevate the standards of AI research by providing researchers and students with feedback from an internationally renowned Program Committee.

AI is a broad research area. Theory, methods, and tools in AI sub-areas such as cognitive science, computational philosophy, computational intelligence, game theory, multi-agent systems, machine learning, multi-agent systems, natural language, representation and reasoning, data mining, speech, computer vision, and the Web. The above methods have broad applications in big data, bioinformatics, biometrics, decision supports, knowledge management, privacy, recommender systems, security, software engineering, spam filtering, surveillance, telecommunications, Web services, and IoT. Submissions received by MIWAI 2019 were wide-ranging and covered both theories as well as applications.

This year's 13th edition of MIWAI was held in Kuala Lumpur, Malaysia, during November 17–19, 2019. This volume contains papers selected for presentation at MIWAI 2019. MIWAI 2019 received 53 full papers from 23 countries including Australia, Bangladesh, Brazil, Brunei Darussalam, China, France, Georgia, Hungary, India, Indonesia, Japan, Malaysia, Nigeria, Pakistan, Philippines, Romania, Slovakia, South Africa, South Korea, Sri Lanka, Thailand, Tunisia, and Vietnam. Following the success of previous MIWAI conferences, MIWAI 2019 continued the tradition of a rigorous review process.

At the end, a total of 25 papers were accepted with an acceptance rate of 47%. Among 25 papers, 19 papers received positive reviews and were accepted as regular papers and the rest were deemed suitable for publication as short papers. Each submission was carefully reviewed by at least two members from a Program Committee consisting of 82 AI experts from 23 countries, some papers received up to five reviews when necessary. The reviewing process was double-blind. Many of the papers that were excluded from the proceedings showed promise but the quality of the proceedings had to be maintained. We would like to thank all authors for their submissions. Without their contribution, this conference would not have been possible.

In addition to the papers published in the proceedings, the technical program included a keynote talk and we thank the keynote speakers for accepting our invitation. We are also thankful to the local organizers in Kuala Lumpur for the excellent hospitality and for making all the necessary arrangements for the conference. Special thanks to Artificial Intelligence Association of Thailand (AiAT) and the Faculty of Informatics, Mahasarakham University for supporting this conference.

We acknowledge the use of the EasyChair conference system for the paper submission, review, and compilation process. Last but not least, our sincere thanks to Alfred Hofmann, Anna Kramer, and the excellent LNCS team at Springer for their support and cooperation in publishing the proceedings as a volume of the *Lecture Notes in Computer Science*.

November 2019   Rapeeporn Chamchong
                                                            Kok Wai Wong

Organization

Steering Committee

Arun Agarwal	University of Hyderabad, India
Rajkumar Buyya	University of Melbourne, Australia
Patrick Doherty	University of Linkoping, Sweden
Rina Dechter	University of California, Irive, USA
Leon Van Der Torre	University of Luxembourg, Luxembourg
Peter Haddawy	Mahidol University, Thailand
Jérôme Lang	University Paris-Dauphine, France
James F. Peters	University of Manitoba, Canada
Somnuk Phon-Amnuaisuk	UTB, Brunei
Srinivasan Ramani	IIIT Bangalore, India
C. Raghavendra Rao	University of Hyderabad, India

Honorary Advisors

Sasitorn Kaewman	Mahasarakham University, Thailand
Huda Ibrahim	Universiti Utara Malaysia, Malaysia

Conveners

Richard Booth	Cardiff University, UK
Chattrakul Sombattheera	Mahasarakham University, Thailand

General Co-chairs

Thanaruk Theeramunkong	SIIT, Thailand
Ku Ruhana Ku Mahmud	Universiti Utara Malaysia, Malaysia

Program Co-chairs

Rapeeporn Chamchong	Mahasarakham University, Thailand
Kok Wai Wong	Murdoch University, Australia

Program Committee

Arun Agarwal	University of Hyderabad, India
Grigoris Antoniou	University of Huddersfield, UK
Adham Atyabi	University of Colorado Colorado Springs and Seattle Children's Research Institute, USA
Thien Wan Au	Universiti Teknologi Brunei, Brunei

Costin Badica	University of Craiova, Romania
Raj Bhatnagar	University of Cincinnati, USA
Richard Booth	Cardiff University, UK
Zied Bouraoui	CRIL-CNRS, Artois University, France
Gauvain Bourgne	CNRS, Sorbonne Université, UPMC Paris 06, LIP6, France
Rapeeporn Chamchong	Mahasarakham University, Thailand
Zhicong Chen	Fuzhou University, China
Suwannit-Chareen Chit	Universiti Utara, Malaysia
Phatthanaphong Chomphuwiset	Mahasarakham University, Thailand
Sook Ling Chua	Multimedia University, Malaysia
Todsanai Chumwatana	Rangsit University, Thailand
Abdollah Dehzangi	Morgan State University, USA
Juergen Dix	Clausthal University of Technology, Germany
Nhat-Quang Doan	University of Science and Technology of Hanoi, Vietnam
Abdelrahman Elfaki	University of Tabuk, Saudi Arabia
Lk Foo	Multimedia University, Malaysia
Hui-Ngo Goh	Multimedia University, Malaysia
Chatklaw Jareanpon	Mahasarakham University, Thailand
Himabindu K.	Vishnu Institute of Technology, Bhimavaram, India
Manasawee Kaenampornpan	Mahasarakham University, Thailand
Ng Keng Hoong	Multimedia University, Malaysia
Kok Chin Khor	Universiti Tunku Abdul Rahman, Malaysia
Suchart Khummanee	Mahasarakham University, Thailand
Ven Jyn Kok	National University of Malaysia, Malaysia
Satish Kolhe	North Maharashtra University Jalgaon, India
Raja Kumar	Taylor's University, Malaysia
Chee Kau Lim	University of Malaya, Malaysia
Chidchanok Lursinsap	AVIC Research Center, Chulalongkorn University, Thailand
Sebastian Moreno	Universidad Adolfo Ibañez, Chile
Sven Naumann	University of Trier, Germany
Atul Negi	University of Hyderabad, India
Thi Phuong Nghiem	USTH, Vietnam
Dung D. Nguyen	Institute of Information Technology, Vietnam Academy of Science and Technology, Vietnam
Thi-Oanh Nguyen	VNU University of Science, Vietnam
Tho Quan	Hochiminh City University of Technology, Vietnam
Srinivasan Ramani	IIIT Bangalore, India
Alexis Robbes	University of Tours, France
Annupan Rodtook	Ramkhamhaeng University, Thailand
Harvey Rosas	University of Valparaiso, Chile

Adrien Rougny	Biotechnology Research Institute for Drug Discovery, National Institute of Advanced Industrial Science and Technology, Japan
Jose H. Saito	Universidade Federal de São Carlos, Brazil
Nicolas Schwind	National Institute of Advanced Industrial Science and Technology (AIST), Japan
Myint Myint Sein	University of Computer Studies, Myanmar
Jun Shen	University of Wollongong, Australia
Guillermo R. Simari	Universidad del Sur in Bahia Blanca, Argentina
Alok Singh	University of Hyderabad, India
Dominik Slezak	University of Warsaw, Poland
Chattrakul Sombattheera	Mahasarakham University, Thailand
Panida Songrum	Mahasarakham University, Thailand
Frieder Stolzenburg	Harz University of Applied Sciences, Germany
Olarik Surinta	Mahasarakham University, Thailand
Ilias Tachmazidis	University of Huddersfield, UK
Jaree Thongkam	Mahasarakham University, Thailand
Thanh-Hai Tran	MICA, Vietnam
Suguru Ueda	Saga University, Japan
Chau Vo	Ho Chi Minh City University of Technology and Vietnam National University – HCMC, Vietnam
Chalee Vorakulpipat	NECTEC, Thailand
Kewen Wang	Griffith University, Australia
Kevin Wong	Murdoch University, Australia

Publicity Chair

Phatthanaphong Chomphuwiset	Mahasarakham University, Thailand

Organizing Co-chairs

Suwannit Chareen Chit	Universiti Utara Malaysia, Malaysia
Manasawee Kaenampornpan	Mahasarakham University, Thailand

Secretary Co-chair

Suwich Tirakoat	Mahasarakham University, Thailand

Web Administration

Olarik Surinta	Mahasarakham University, Thailand
Panich Sudkhot	Mahasarakham University, Thailand

Additional Reviewers

Na Li
Nita Patil
Srinivasan Ramani
Mohamed Rym
Snehalata Shirude
Sophie Siebert
Kai Steckhan
Ma Truong Thanh

Algorithms to Find Interesting and Interpretable High Utility Patterns in Symbolic Data (Keynote Abstract)

Philippe Fournier-Viger

Harbin Institute of Technology, Shenzhen, China

Abstract. Discovering interesting and useful patterns in symbolic data has been the subject of numerous studies. It consists of extracting patterns from data that meet a set of requirements specified by a user. Although early research work in this domain have mainly focused on identifying frequent patterns (e.g. itemsets), nowadays many other types of interesting patterns have been proposed and more complex data types and pattern types are considered. Mining patterns have applications in many fields as they provide glass-box models that are generally easily interpretable by humans either to understand the data or support decision-making. This talk will first highlight limitations of early work on frequent pattern mining and provide an overview of state-of-the-art problems and techniques related to identifying interesting patterns in symbolic data. Topics that will be discussed include high utility patterns, locally interesting patterns, periodic patterns, and statistically significant patterns. Lastly, the SPMF open-source software will be mentioned and opportunities related to the combination of pattern mining techniques with traditional artificial intelligence techniques.

Contents

Short Papers

Regular Papers

Text Relation Extraction Using Sentence-Relation Semantic Similarity

Mohamed Lubani[(✉)] and Shahrul Azman Mohd Noah

Center for Artificial Intelligence Technology, Faculty of Information Science
and Technology, Universiti Kebangsaan Malaysia,
43600 Bangi, Selangor, Malaysia
mohamed.lubani@siswa.ukm.edu.my, shahrul@ukm.edu.my

Abstract. There is a huge amount of available information stored in unstructured plain text. Relation Extraction (RE) is an important task in the process of converting unstructured resources into machine-readable format. RE is usually considered as a classification problem where a set of features are extracted from the training sentences and thereafter passed to a classifier to predict the relation labels. Existing methods either manually design these features or automatically build them by means of deep neural networks. However, in many cases these features are general and do not accurately reflect the properties of the input sentences. In addition, these features are only built for the input sentences with no regard to the features of the target relations. In this paper, we follow a different approach to perform the RE task. We propose an extended autoencoder model to automatically build vector representations for sentences and relations from their distinctive features. The built vectors are high abstract continuous vector representations (embeddings) where task related features are preserved and noisy irrelevant features are eliminated. Similarity measures are then used to find the sentence-relation semantic similarities using their representations in order to label sentences with the most similar relations. The conducted experiments show that the proposed model is effective in labeling new sentences with their correct semantic relations.

Keywords: Embeddings · Natural language processing · Neural networks · Relation extraction

1 Introduction

In natural language processing (NLP), relation extraction (RE) is the task of identifying the semantic relations between concepts in natural language text. Binary relation extraction is the most common type of RE where semantic relations between two entities/nominals are extracted from natural language sentences. For example, *CapitalOf* (Paris, France) and *LocatedAt*(Eiffel Tower, Paris) are examples of binary relations between two entities, whereas *ProducedBy*(Honey, Bee) is an example of a binary relation between two nominals. Semantic relations can also be identified between more than two entities. Such high-order relations can be extracted by analysing and combining the binary relations between pairs of entities. RE is an important part of information extraction (IE) where structured knowledge is extracted from unstructured raw text.

© Springer Nature Switzerland AG 2019
R. Chamchong and K. W. Wong (Eds.): MIWAI 2019, LNAI 11909, pp. 3–14, 2019.
https://doi.org/10.1007/978-3-030-33709-4_1

Thus, RE can be used in many NLP tasks, such as ontology learning and population [1], question answering [2], and information retrieval [3].

Many existing methods consider RE as a classification task. Such methods extract features from the input sentences and use them to train a classifier to predict the target relation labels. In this paper, RE is performed from text based on sentence-relation semantic similarities. A model that extends the concept of autoencoders is proposed to map the representations of sentences and their target relations to close points in the same vector space. The proposed model automatically builds high abstract continuous vector representations by enhancing distinctive features and eliminating noisy irrelevant ones. Sentences are then assigned the labels of their most similar relations using direct similarity measure calculations.

The remaining sections of this paper are organised as follows. Section 2 provides details about various relation extraction approaches. The proposed model to build the vector representations is described in Sect. 3. The conducted experiments and the collected results are discussed in Sect. 4. Finally, Sect. 5 concludes the paper and presents possible future extensions.

2   Literature Review

Various methods have been proposed to perform text relation extraction. The bootstrapping methods or rote extractors perform the RE task using extraction patterns. These methods exploit the duality between a set of seed instances and the extraction patterns. For example, Brin [4], Snowball [5], and Ravichandran et al. [6] used a corpus to locate occurrences of a given set of seed instances to generate the extraction patterns. The generated patterns were then used to extract new instances and the process was repeated. The bootstrapping methods generate a large number of extraction patterns at the cost of accumulating errors from noisy patterns, i.e. the semantic drift problem. To improve the precision of the extraction patterns, Alfonseca et al. [7, 8] described a method to estimate the precision of rote extractor patterns. KnowItAll [9] employed a Naïve Bayes classifier to evaluate the extracted relations and filter noisy extractions.

The most popular approach for semantic relation extraction is the supervised approach, where the relation extraction is treated as a classification problem. Methods of this approach are either directly supervised by a tagged corpus or distantly supervised by a knowledge base using the distant supervision hypothesis [10]. Syntactic and semantic features are extracted from the positive relation examples and fed to a machine learning classifier to learn the correct relation classifications. Based on the method used to build a set of context features, supervised methods can be divided into three groups. The first group includes the methods that manually design the set of features to be used to train the classification model [11–15]. These methods define both syntactic features (e.g. the actual words, the part-of-speech (POS) tags, and the entity types) and semantic features (e.g. the path between the entities in the dependency tree). The second group contains the kernel-based methods that implicitly utilise the structural features of sentences. These methods can be subsequence-based that use the sequence of words [16] or tree-based that use parse trees and dependency trees [17–21]. The third group contains the methods that utilise neural network architectures to automatically extract features from the input such as the deep learning methods [12–14]. To enhance the ability of the neural network to capture useful semantic features for relation extraction, Su et al. [22] proposed a Recurrent Neural Network (RNN) model that employs the global statistics of relations in a large corpus. To capture the structural features between entities, a Piecewise Convolutional Neural Network (PCNN) model is proposed in [23] to obtain the features of various sentence parts based on the position of the two entities. A dependency-based neural network model is proposed in [24] to capture the features of the dependency paths and dependency sub-trees of the input sentences. The built features in all the previously mentioned groups are passed to a classifier such as a Support Vector Machine (SVM) or to an additional softmax layer on top of the neural network to provide the final relation predictions. Relation representations are used in [25] instead of exact relation labels to train the model. The method uses the relation representations previously generated by the TransE model [26] to supervise the training of the model.[1] The model in [25] uses a PCNN [23] to obtain the encodings of the input sentences to be passed to a linear layer to predict the target relation representations.

Methods that automatically extract the features can be seen as sentence encoders, where high abstract representations are built for the input sentences. However, most methods do not use these representations outside the scope of the neural network model. The representations are only built to be used by the classification layer to label the sentences with their corresponding relation labels. In addition, most methods only build and optimise sentence representations during the learning process of the model with no regard to the target relation representations. This paper aims to generate the representations of sentences and their target relations from their distinctive features to be used to find the sentence-relation semantic similarities for the task of relation extraction. The proposed model extends the concept of autoencoders to map sentences and their target relations to close points in the same vector space. Unlike previous approaches such as in [25], the proposed training process optimises both the sentences and relations representations to only keep distinctive features that are helpful for the task of relation extraction and remove noisy irrelevant ones. The built representations are then used to assess the sentence-relation semantic relatedness and perform RE based on similarity measures instead of machine learning classifiers.

3   Building Vector Representations

This section describes the proposed method to build the vector representations for sentences and their target relations. The following sections include details about the structure of the proposed model, the initial input representation, and the training objective.

---

[1] TransE provides continuous vector representations for entities and their relations in a knowledge graph.

3.1   The Model

The proposed model aims to map the representations of sentences and their corresponding relations to close points in the same vector space. Once the mapping is completed, relation extraction can be performed by assessing the sentence-relation semantic similarities using similarity measures. The model learns high abstract representations for the input sentences and any new sentences that express the same semantic relation between two entities. The proposed model is built based on the concept of autoencoders. Autoencoders are unsupervised neural networks first proposed in [27] to learn internal representations of the input vectors. An autoencoder consists of two parts. The first part is the encoder part that encodes the input vectors into dense representations with smaller dimensions than the input vectors.[2] The second part is the decoder that attempts to reconstruct the input vectors from their dense representations. Thus, the autoencoder is trained solely using the input vectors as input and output without the need for any labels. Both the encoder and decoder parts are neural networks that produce numerical output vectors for their inputs. These neural networks may consist of multiple consecutive hidden layers. The result of the encoder/decoder is the output of the last hidden layer in the corresponding neural network. The output of a hidden layer y_j is defined as follows:

$$y_j = f\left(W_{ij}v + B_j\right) \tag{1}$$

where W_{ij} is the weight matrix between the ith and the jth consecutive layers, v is the network's input vector if the jth layer is the first hidden layer, otherwise it is the output of the previous hidden layer. f is an activation function and B_j is a bias vector. The values of W_{ij} and B_j will be adjusted during the training process.

For a set of n input vectors, the training of an autoencoder is performed by minimising the following objective function:

$$O = \sum_{i=1}^{n} ||v_i - dec(enc(v_i))||^2 \tag{2}$$

where v_i is the ith input vector in the training set, while $enc(v)$ and $dec(v)$ are the results of the encoder and decoder parts respectively.

If the reconstruction process is achieved with small errors, the dense vector representations calculated by the encoder part are seen as representations that hold the distinctive features needed to reconstruct the input vectors. Therefore, autoencoders build high abstract vector representations of their input by keeping the most distinctive features and removing the noisy ones that are not helpful in the reconstruction phase. The decoder part is only used to train the autoencoder. Once trained, only the encoder part will be used to find high abstract representations of new input vectors. The result of the encoder part can be passed as input to train another autoencoder and so on, which

---

[2] Autoencoders that produce representations smaller that the input vectors are called undercomplete autoencoders, which are used to learn meaningful features and prevent copying the input vectors.

allows for building more abstract representations of the input. A stacked autoencoders architecture is then constructed by stacking the encoder parts of multiple pre-trained autoencoders. Such stacked architectures have two main benefits. The first benefit is that they build high abstract and compact representations of the input vectors, where only distinctive features are preserved and noisy irrelevant ones are eliminated. The second benefit is the continuous training of consecutive autoencoders. Training individual autoencoders is more likely to suffer from the local minima problem.[3] Stacked autoencoders structure helps to reach a global minimum by continuous training, where subsequent autoencoders are trained based on the knowledge learned by the previous autoencoders.

The architecture of the proposed model is stacked autoencoders built using the encoder parts of two separately pre-trained autoencoders. Each of the pre-trained autoencoders consists of a neural network with two hidden layers. Figure 1 shows the architecture of the proposed model.

As shown in Fig. 1, only the encoder parts of the two autoencoders are used to construct the final model. Since each autoencoder consists of two hidden layers, the result of the encoder part of each of the two used autoencoders can be defined as the following:

$$enc^{(i)}(v) = \sigma\left(W_2^{(i)}\left(\sigma\left(W_1^{(i)}v + B_1^{(i)}\right)\right) + B_2^{(i)}\right) \tag{3}$$

Where $W_1^{(i)}$ and $W_2^{(i)}$ are the weight matrixes of the two hidden layers of the ith autoencoder, $B_1^{(i)}$ and $B_2^{(i)}$ are the bias vectors, and $\sigma(x) = \frac{1}{1 + \exp(-X)}$ is the sigmoid function.

The final output of the stacked autoencoders model is defined as follows:

$$enc(v) = enc^{(2)}\left(enc^{(1)}(v)\right) \tag{4}$$

Where $enc^{(1)}$ and $enc^{(2)}$ are the encoding results of the first and second autoencoders, respectively.

3.2 The Initial Input Representations

Since the used model is a neural network, the input should be converted to numeric vectors before passing it to the model to be processed. The model is supervised by a set of training sentences labelled with semantic relations. Each sentence contains two named entities and is labelled with the semantic relation that is expressed between the two entities. Sentences in the training set are pre-processed to detect and label the two named entities. A syntactic dependency parser is then used to build the dependency tree for each training sentence. A dependency parser examines the grammatical structure of natural language text and identifies syntactic dependency relations between words.

---

[3] This is a common problem in gradient descent optimisation where the error minimisation stops before reaching the global optimised solution.

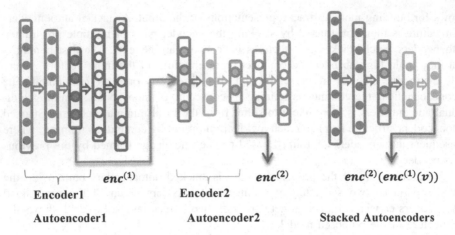

Fig. 1. The architecture of the proposed model

After building the dependency parse tree for the input sentences, the shortest dependency path between the two entities is extracted as the minimum sequence of words that connects the two entities in the dependency tree. For example, in the sentence "*Paris is the most populous city of France*", the shortest dependency path between 'Paris' and 'France' is the sequence ['is', 'city', 'of']. Pre-trained embeddings are used to represent words in the shortest dependency path. Embeddings are continuous vector representations that signify the syntactic and semantic properties of words from their occurrences in a large corpus. Embeddings can be built using either predictive models such as word2vec [28] or count-based models such as GloVe [29]. The final representations of the input sentences are built by averaging the embeddings of individual words in their shortest dependency path sequences.

Embeddings are also used to represent semantic relations by averaging the embeddings of individual tokens in their labels. For example, the relation "cause-effect" is represented as the average of the embeddings of both words, i.e. "cause" and "effect". The initial relation representations have the same distribution as the initial sentence representations since both are built from the same pre-trained embeddings. These initial representations are passed as input to the proposed model in order to build the final compact vector representations.

3.3 The Training Objective

This paper extends the concept of autoencoders to map the input vectors to their intended output vectors, which are not necessarily the same as the input vectors. The goal is to map the sentence representations and the representations of their target semantic relations to close points in the same vector space. As previously explained, the proposed model is composed of two stacked autoencoders. Each autoencoder is trained separately to reconstruct the vector representation of the target relation from the representations of the input sentences. This makes the encoder representations, i.e. the compact representations of all the sentences labelled with the same target relation, close to each other since these representations are used to reconstruct the same output. To make these compact representations close to the encoder representation of the target relation, each autoencoder is also trained to minimise the reconstruction error of the target relation using its own representation as input. In other words, the autoencoder is trained to predict the relation representation from the representations of its sentences as well as predict the relation representation from itself. This makes the compact representations of the sentences and their target relation become close to each other in the same vector space. To restrict the generalisation, each autoencoder is also trained to map the representations of noise sentences to a random relation representation. The noise sentences are negative examples of the target relation for which the autoencoder is being trained. This restricts the generalisation capabilities of the autoencoder and prevents pointing any sentence from the same distribution as the positive training sentence to the target relation. By extending the objective in Eq. (2), the training of each autoencoder is achieved by minimising the following objective function:

$$\mathcal{F} = \sum_{i=1}^{n} ||r - dec(enc(v_i))||^2 + ||r - dec(enc(r))||^2 + ||r' - dec(enc(v_i'))||^2 \quad (5)$$

where n is the number of training sentences for the target relation that has the initial representation, r, v_i is the initial representation of the ith sentence in the training set, v_i' is the representation of a randomly selected negative (noise) sentence, and r' is the representation of a random relation.

When training the second autoencoder, the vectors r, v_i, v_i', and r' are first encoded using the encoder of the first autoencoder and then used to train the second autoencoder based on the objective in Eq. (5). Once both autoencoders are trained, the final model is constructed by stacking the encoder parts of the individual autoencoders as shown in Fig. 1. For relation extraction, the input representations of new test sentences are passed to the model to generate their encodings. These encodings are then compared with the encoding of the target relation using a similarity measure. The sentences are labelled with the target relation if their similarities are higher than the average of all similarities of the training sentences with the target relation.[4]

4 Experiments and Results

This section discusses the implementation details and presents the results of the conducted experiments to evaluate the proposed model. SpaCy dependency parsing [30] was used to generate the parse trees for the input sentences. Once generated, these trees were converted into graphs to find the shortest paths between two nodes, i.e. the two entities in each sentence. SpaCy embeddings that were built based on the GloVe algorithm were used to represent individual words and build the input vector representations for sentences and relations. To implement the proposed model, the present

---

[4] For better generalisation from the training to the testing sentences, the threshold may be reduced by a small fixed value for all the semantic relations.

study utilised Google's Tensorflow library [31]. Tensorflow library was used to build and train the two autoencoders in the model. The input layer of the first autoencoder had a size of 300, which is the same size as the embedding vectors produced by SpaCy. With two hidden layers and an undercomplete property, the encoder network of the first autoencoder had the structure of (300-150-75), where 75 is the size of the encoder output. The decoder network attempted to reconstruct the input representations from the results of the encoder network; thus had the structure of (75-150-300). The second autoencoder took the encoded results of the first autoencoder as input and had the structure of (75-50-25-50-75). The final model was obtained by stacking the encoder parts of the two pre-trained autoencoders and thus had the structure of (300-150-75-50-25), where 25 is the size of the encoded output of the proposed model. Cosine similarity was employed to calculate the similarities between the encodings of sentences and the target relations in order to assign relation labels to sentences. Cosine similarity is chosen due to its ability to reflect the directional similarities between the encodings.

To test the proposed model, SemEval 2007-task4 dataset [32] was used. The dataset contains training and testing examples for seven semantic relations: Cause-Effect, Instrument-Agency, Product-Producer, Origin-Entity, Theme-Tool, Part-Whole, and Content-Container. Each relation has a training file with 140 training sentences and a testing file with about 70 testing sentences. The two entities in each sentence in the dataset are already detected and labelled as e1 and e2, respectively. Training and testing sentences are labelled as either true or false to indicate the existence or absence of the semantic relation between the two entities. In the experiments, only the true labelled sentences from each file were considered. The proposed model was trained for each semantic relation separately using the true labelled sentences from its training file. The true sentences of other semantic relations were considered as the source for the noise sentences while training the model for the target relation. The false labelled sentences could also be used as a source for the noise sentences for the target relation. However, using the sentences of other relations provides a wider spectrum of noise sentences distribution. For evaluation, the following terms were defined. True Positive (TP) is the number of correctly classified true labelled sentences from the testing file of the target relation. False Negative (FN) is the number of misclassified true labelled sentences from the testing file of the target relation. False Positive (FP) is the number of true labelled sentences of other semantic relations that are misclassified as the target relation. To evaluate the model, the following evaluation measures were used. The precision (P) to indicate the quality of the model was defined as $P = TP/(TP + FP)$. The recall (R) to indicate the completeness, i.e. the coverage of the model was defined as $R = TP/(TP + FN)$. The F-measure was used to combine both the precision and the recall and was defined as $F = 2 \times P \times R/(P + R)$. Table 1 shows the values of these measures for the proposed model for each of the seven semantic relations in the used dataset. The table also shows the baseline F-measure (alltrue) value, i.e. Base-F, for each relation as reported in [32].

As displayed in Table 1, the proposed model had different precision and recall values for different semantic relations. The 'Theme-Tool' relation had the highest precision, whereas the 'Content-Container' relation had the highest recall. The average values for the proposed model over all relations were 73.7% and 67.0% for precision

Table 1. The results of the proposed model for each semantic relation

The relation	Proposed model			Baseline
	Precision	Recall	F-measure	F-measure
Cause-Effect	75.6	77.5	76.5	67.8
Instrument-Agency	88.0	57.8	69.8	65.5
Product-Producer	70.4	80.6	75.1	80.0
Origin-Entity	33.9	52.7	41.3	61.5
Theme-Tool	95.2	68.9	80.0	58.0
Part-Whole	59.0	50.0	54.1	53.1
Content-Container	93.9	81.5	87.3	67.9

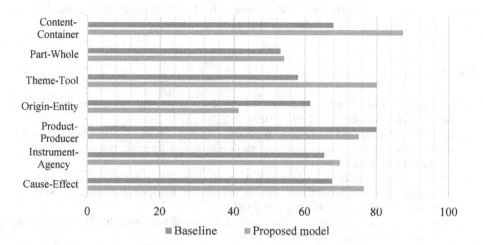

Fig. 2. The F-measure values of the proposed model and the baseline for each semantic relation

and recall, respectively. Figure 2 graphically illustrates the F-measure values of the proposed model and the baseline for each of the seven semantic relations.

As Fig. 2 shows, the proposed model had higher F-measure values than the baseline for all semantic relations except 'Origin-Entity' and 'Product-Producer'. This could be due to the semantic nature of these two relations, which makes their sentences similar in structure to the sentences of other relations. The highest F-measure value for the proposed model was obtained for the 'Content-Container' relation at 87.3%, which was significantly higher than the corresponding baseline value. Considering the average values, the model had an average F-measure value of 69.1% as compared to 64.8% average baseline values for all semantic relations.

5 Conclusion

In this paper, a model was proposed to build the vector representations of sentences and their target relations to be used to find sentence-relation semantic similarities for the task of relation extraction. The model builds vectors that hold distinctive features that can be used to assign semantic relations to new sentences based on the similarities between their corresponding vectors.

The built representations can be utilised to find the similarities between any sentence/relation pair such as sentence-sentence similarities and relation-relation similarities. These representations can also be built for other languages such as Malay language to measure sentence-sentence semantic similarities [33, 34]. For future work, it is recommended to investigate the benefits of using these similarities for the task of relation extraction. Additionally, future research should investigate and implement solutions to determine the direction of the semantic relations.

References

1. Lubani, M., Noah, S.A.M., Mahmud, R.: Ontology population: approaches and design aspects. J. Inf. Sci. **45**(4), 502–515 (2019)
2. Hazrina, S., Sharef, N.M., Ibrahim, H., Murad, M.A.A., Noah, S.A.M.: Review on the advancements of disambiguation in semantic question answering system. Inf. Process. Manage. **53**(1), 52–69 (2017)
3. Nasution, M.K., Noah, S.A.: Information retrieval model: a social network extraction perspective. In International Conference on Information Retrieval & Knowledge Management, pp. 322–326, IEEE, Kuala Lumpur (2012)
4. Brin, S.: Extracting patterns and relations from the world wide web. In: The World Wide Web and Databases, pp. 172–183 (1998)
5. Agichtein, E., Gravano, L.: Snowball: extracting relations from large plain-text collections. In: Proceedings of the Fifth ACM Conference on Digital Libraries, pp. 85–94. ACM (2000)
6. Ravichandran, D., Hovy, E.: Learning surface text patterns for a question answering system. In: Proceedings of the 40th Annual Meeting of the Association for Computational Linguistics (ACL), Philadelphia, pp. 41–47 (2002)
7. Alfonseca, E., Ruiz-Casado, M., Okumura, M., Castells, P.: Towards large-scale non-taxonomic relation extraction: estimating the precision of rote extractors. In: The 2nd Workshop on Ontology Learning and Population, Sydney, Australia, pp. 49–56 (2006)
8. Alfonseca, E., Castells, P., Okumura, M., Ruiz-Casado, M.: A rote extractor with edit distance-based generalisation and multi-corpora precision calculation. In: Proceedings of the COLING/ACL on Main Conference Poster Sessions, pp. 9–16. Association for Computational Linguistics (2006)
9. Etzioni, O., et al.: Unsupervised named-entity extraction from the web: an experimental study. Artif. Intell. **165**, 91–134 (2005)
10. GuoDong, Z., Jian, S., Jie, Z., Min, Z.: Exploring various knowledge in relation extraction. In: Proceedings of the 43rd Annual Meeting of the ACL, pp. 427–434. Association for Computational Linguistics, Ann Arbor (2005)
11. Chen, Y., Li, W., Liu, Y., Zheng, D., Zhao, T.: Exploring deep belief network for Chinese relation extraction. In: Proceedings of the CIPS-SIGHAN Joint Conference on Chinese Language Processing (CLP 2010), Beijing, China (2010)

12. Liu, C., Sun, W., Chao, W., Che, W.: Convolution neural network for relation extraction. In: Motoda, H., Wu, Z., Cao, L., Zaiane, O., Yao, M., Wang, W. (eds.) ADMA 2013. LNCS (LNAI), vol. 8347, pp. 231–242. Springer, Heidelberg (2013). https://doi.org/10.1007/978-3-642-53917-6_21

13. Zeng, D., Liu, K., Lai, S., Zhou, G., Zhao, J.: Relation classification via convolutional deep neural network. In: International Conference on Computational Linguistics, Dublin, Ireland, pp. 2335–2344 (2014)

14. Suchanek, F.M., Ifrim, G., Weikum, G.: LEILA: learning to extract information by linguistic analysis. In: Proceedings of the 2nd Workshop on Ontology Learning and Population: Bridging the Gap between Text and Knowledge, Sydney, Australia, pp. 18–25 (2006)

15. Zelenko, D., Aone, C., Richardella, A.: Kernel methods for relation extraction. J. Mach. Learn. Res. 3, 1083–1106 (2003)

16. Culotta, A., Sorensen, J.: Dependency tree kernels for relation extraction. In: ACL 2004 Proceedings of the 42nd Annual Meeting on Association for Computational Linguistics, p. 423. Association for Computational Linguistics, Barcelona (2004)

17. Bunescu, R., Mooney, R.: A shortest path dependency kernel for relation extraction. In: Proceedings of Human Language Technology Conference and Conference on Empirical Methods in Natural Language Processing, pp. 724–731. Association for Computational Linguistic, Vancouver (2005)

18. Zhou, G., Zhang, M., Ji, D.H., Zhu, Q.: Tree kernel-based relation extraction with context-sensitive structured parse tree information. In: Proceedings of the 2007 Joint Conference on Empirical Methods in Natural Language Processing and Computational Natural Language Learning (EMNLP-CoNLL), pp. 728–736. Association for Computational Linguistics, Prague (2007)

19. Zhang, M., Zhang, J., Su, J., Zhou, G.: A composite kernel to extract relations between entities with both flat and structured features. In: Proceedings of the 21st International Conference on Computational Linguistics and the 44th Annual Meeting of the Association for Computational Linguistics, pp. 825–832. Association for Computational Linguistics, Sydney, Australia (2006)

20. Bunescu, R.C., Mooney, R.J.: Subsequence kernels for relation extraction. In: NIPS 2005 Proceedings of the 18th International Conference on Neural Information Processing Systems, pp. 171–178. MIT Press, Cambridge (2005)

21. Mintz, M., Bills, S., Snow, R., Jurafsky, D.: Distant supervision for relation extraction without labeled data. In: Proceedings of the 47th Annual Meeting of the ACL and the 4th IJCNLP of the AFNLP, Suntec, Singapore, pp. 1003–1011 (2009)

22. Su, Y., Liu, H., Yavuz, S., Gür, I., Sun, H., Yan, X.: Global relation embedding for relation extraction. In: Proceedings of the 2018 Conference of the North American Chapter of the Association for Computational Linguistics: Human Language Technologies, Volume 1 (Long Papers), pp. 820–830. Association for Computational Linguistics, New Orleans (2018)

23. Zeng, D., Liu, K., Chen, Y., Zhao, J.: Distant supervision for relation extraction via piecewise convolutional neural networks. In: Proceedings of the 2015 Conference on Empirical Methods in Natural Language Processing, pp. 1753–1762. Association for Computational Linguistics, Lisbon (2015)

24. Liu, Y., Li, S., Wei, F., Ji, H.: Relation classification via modeling augmented dependency paths. IEEE/ACM Trans. Audio Speech Lang. Process. 24(9), 1589–1598 (2016)

25. Bordes, A., Usunier, N., Garcia-Durán, A., Weston, J., Yakhnenko, O.: Translating embeddings for modeling multi-relational data. In: NIPS 2013 Proceedings of the 26th International Conference on Neural Information Processing Systems, vol. 2, pp. 2787–2795. Curran Associates Inc., Lake Tahoe (2013)

26. Wang, G., et al.: Label-free distant supervision for relation extraction via knowledge graph embedding. In: Proceedings of the 2018 Conference on Empirical Methods in Natural Language Processing, pp. 2246–2255. Association for Computational Linguistics, Brussels (2018)

27. Rumelhart, D.E., McClelland, J.L., Asanuma, C.: Learning internal representations by error propagation. In: Parallel Distributed Processing: Explorations in the Microstructure of Cognition, pp. 318–362. MIT Press Cambridge (1986)

28. Mikolov, T., Sutskever, I., Chen, K., Corrado, G., Dean, J.: Distributed representations of words and phrases and their compositionality. In: Advances in Neural Information Processing Systems, pp. 3111–3119. Curran Associates Inc., USA (2013)

29. Pennington, J., Socher, R., Manning, C.: GloVe: global vectors for word representation. In: Proceedings of the 2014 Conference on Empirical Methods in Natural Language Processing (EMNLP), pp. 1532–1543. Association for Computational Linguistics, Doha (2014)

30. Abadi, M., et al.: Tensorflow: Large-scale machine learning on heterogeneous distributed systems. arXiv preprint arXiv:1603.04467 (2016)

31. Goldberg, Y., Nivre, J.: A dynamic Oracle for arc-eager dependency parsing. In: Proceedings of COLING 2012, pp. 959–976. The COLING 2012 Organizing Committee, Mumbai, India (2012)

32. Girju, R., Nakov, P., Nastase, V., Szpakowicz, S., Turney, P., Yuret, D.: SemEval-2007 task 04: classification of semantic relations between nominals. In: SemEval 2007 Proceedings of the 4th International Workshop on Semantic Evaluations, pp. 13–18. Association for Computational Linguistics, Prague (2007)

33. Noah, S.A., Omar, N., Amruddin, A.Y.: Evaluation of lexical-based approaches to the semantic similarity of Malay sentences. J. Quantit. Ling. 22(2), 135–156 (2015)

34. Noah, S.A., Amruddin, A.Y., Omar, N.: Semantic Similarity Measures for Malay Sentences. In: Goh, D.H.-L., Cao, T.H., Sølvberg, I.T., Rasmussen, E. (eds.) ICADL 2007. LNCS, vol. 4822, pp. 117–126. Springer, Heidelberg (2007). https://doi.org/10.1007/978-3-540-77094-7_19

Internet of Things Sensors and Actuators Layered Fog Service Delivery Model SALFSD

Abdulsalam Alammari[✉], Salman Abdul Moiz, and Atul Negi

School of Computer and Information Sciences, University of Hyderabad,
Hyderabad, India
{alammari,salman}@uohyd.ac.in, atulcs@uohyd.ernet.in

Abstract. Internet of Things (IoT) is being widely adopted for building many domain applications. IoT Sensors and Actuators "things" are the most important and essential components upon which these applications are built. For new businesses, investing in deployment of their own sensors and actuators may be infeasible. Thus it was suggested to have sensing and actuating as a service and as a service delivery model. These services may be free (as in use without charge) or they may have a cost structure (pay as you use). In this work we study the adoption of layered fog architecture to enhance sensing and actuating as a service delivery model using this proposed architecture model. Our main goal to provide multiple sensors and actuators composition options, with high data filtering rates and reduced response time. We propose a layered architecture with built-in fault resistance. We explain the various in context of the literature.

Keywords: Internet of Things · Sensing and Actuating as a Service · Cloud · Fog · Sensors and Actuators

1 Introduction

The increasing amount of objects connected to the internet using various preexisting technologies for communication, storage, ubiquitous and pervasive computing, wireless sensor networks etc, they are now considered as the Internet of Things(IoT). IoT generally uses the existing technologies and paradigms to serve or change the way services are provided to many application domains like smart home, smart grid, smart agriculture, smart forest, healthcare etc.

IoT enables such physical objects to communicate, coordinate and share their data by employing the above mentioned technologies. This arrangement is more suitable for these physical objects and allows to cope with their constraints and limitations to make them smart, rather than as traditional passive objects [1]. In other words, IoT is the enabler connecting the physical devices to the digital world, making numerous physical devices to be connected to the internet and allow remote control heterogeneous devices collaboration.

© Springer Nature Switzerland AG 2019
R. Chamchong and K. W. Wong (Eds.): MIWAI 2019, LNAI 11909, pp. 15–25, 2019.
https://doi.org/10.1007/978-3-030-33709-4_2

IoT devices have well known limitations and constraints (e.g., limited storage, limited processing and communication capabilities) which are mainly being overcome with the help of cloud paradigm that acts as a backbone for remote processing and storage of IoT devices data. Cloud provides IoT with virtually unlimited resources [2]. Besides, sensors as essential components of IoT applications create a large amount of data (Big Data). Big Data has its associated challenges like storage and communication that IoT constrained devices cannot address without the great solutions provided by IoT integration with Cloud. This integration supports Storage as a Service [3], Software as a Service and Platform as a Service [4]. However, the more sensors are connected, more data is sent to cloud leading to increasing transmission overhead and latency issues [5].

Therefore, the questions like, "what to process", and "where to process", or "where to take a decision based on realtime date streaming" are resolved in a more recent paradigm called Fog computing or mini-cloud. Fog computing is mainly about bringing the processing closer to the data sources. However, what processing is to be pulled out from the cloud, and where to be placed in the network topology are the main important issues in defining exactly a fog node and its location [6]. In Fog computing some tasks are offloaded from the cloud to the fog node that significantly reduces the processing time and communication overhead. This results in better performance than having all tasks executed centrally at cloud, such performance enhancement is significant in domains like healthcare [7].

For some IoT application and IoT services providers, investing for their own infrastructure or platform may be costly or infeasible hence it is advantageous to make use facilities provided by cloud like Infrastructure as a Service, Storage as a Service, Software as a Service and Platform as a Service.

It is interesting to note that as of now not much exists in the literature, however new frameworks emerged like Sensing as a Service for Internet of Things [8], Sensing and Actuation as a Service [9], and Sensing and Actuation as a Service Delivery Model [10]. The most important factor in Sensing and Actuating as a Services, from our point of view, is best utilizing the network topology in a way that observation processing, physical nodes selection, virtualization, and realtime decision making etc. to be made in the correct layer for better enhancement of performance.

This work aims at leveraging the layered style (Things and gateway layer, Fog layer, and Cloud layer) for designing a framework for Sensing and Actuating as a Service Delivery for Internet of Things [10]. We call our approach Sensors and Actuators Layered Fog Service Delivery Model (SALFSD). We expect that having fog between the gateway and the cloud will be a good option for such models as it will enhance the performance, reduce the latency, and provide more options for composing the requested types of sensors and actuators belonging to different gateways under the area of end user interest.

The rest of this paper is organised as following. Section two elaborates some preliminary concepts and discusses the related work. Our proposed architecture is then explained in section three. This is followed with a comparison of our work with related work and conclusion in section four and five respectively.

2 Background and Related Work

The idea of Sensing and Actuating as a Service SAaaS was encouraged by the following facts:

- The globally distributed and wide spread of heterogeneous smarts devices.
- Smart devices and sensors and actuators that can be categorised location wise or by their type/functionality.
- The spread of Sensing/Infrastructure/Storage as a service.

Such factors made a robust base for SAaaA.

Sensing and Actuating as a Service provides various types of sensors and actuators resources from sensor networks and personal mobile devices to the end user as services. The end user can then build their own application based on rented composed/aggregated types of resources of their desired locations, functionalities and time. SAaaS also provides the end user with the ability to trigger actuating commands as per their analysis of the scenarios/real/historical data etc.

In SAaaS owners (being individuals, enterprise, universities, etc.) of sensing/actuating devices may contribute their devices services for free or provides the same on rent bases. Devices owners may contribute standalone Sensors/Actuators (S/A) devices or WSN in which they are called contributing nodes which here (whether have single or group of sensors or actuators)must belong to the same administrative domain [9]. End users (the once will take the benefits of the sensors and actuators) might also be individuals, enterprise, universities, etc.

The SAaaS provider acts as the mediator between the two parties and manages all the aspects of the paradigm. Among SAaaS provider responsibilities are devices owners and end user enrolment, Service Level Agreement creation, physical nodes selection, rent calculation, virtualization, etc. SAaaS virtualization has two types. First is the virtualization of the observation data in witch the provider receives the sensors observations and saves them to provide each end user with the required data at their prespecified time. Second is the actuator virtualization where the provider facilitates the end user access (sending an actuator command) to the previously rented actuator. This concludes that the provider must create a virtual set of desired sensors and actuator for each end used as per their requirement. Below is a review of literature that motivated the features of the proposed approach in this paper.

In their work in [9], authors identified the main concepts and actors of SAaaS and proposed an architecture of SAaaS in which they make a Hyervisor in the nodes to abstract the, management and virtualization of sensors and actuators as a virtual instances in the cloud. Even though the Hypervisor works at device level (device level virtualization and abstraction), it enables direct communication with sensing and actuating devices through the Automatic Enforcer which is located between the VolunteerCloud Manager and the Hypervisor. In [11], authors used their hypervisor propesed in [9] to introduce more detailed architecture for SAaaS giving more concentration for management, abstraction and virtualization of sensing resources. Their proposed architecture allows devices to provide their sensors as virtual instance which makes a particular physical device handles concurrent request. Besides, the architecture can compose/aggregate a network of resource instances out of simple individual instances. It worth mentioning here, since the hypervisor is the base of this architecture, that this architecture is device-oriented approach but still concurrent requests of the same physical device are ensured.

Further extension to their works in [9] and [11], authors proposed utility framework [12] for IoT SAaaS approach inside IoT-A reference architecture. This work implemented the earlier proposed idea (implementation for Android mobile) in a real life IoT scenario to show its feasibility.

Stack4Things [13] adopted the OpenStack [14] (Infrastructure as a Service framework) to propose a framework for Sensing and Actuating as a Service (device-centric approach). Stack4Things shows the detailed subsystems for resources management and their observation data, and some use cases were demonstrated. It is worth mentioning that the above mentioned related works follow the traditional device-cloud fashion.

A cloud edge-centric Sensing and Actuating as a Service Delivery Model (SAaaSDM) was introduced in [10] where end users are offered sensed, actuated and computed data from existing resources belonging to different owners. The resources are given to the end users on rent bases. SAaaSDM introduced participatory node, virtual node, and review management components in its architecture. Therefore, the model aims at providing robust management for the existing IoT sensors and actuators in Cloud Edge-Centric fashion and process/filter their data in the Edge Node(Cloud gateway).

3   SALFSD Architecture

In this paper we follow the a layered topology for Sensing and Actuating as Service Delivery Model SAaaSDM which was introduced in [10] adopting Cloud Edge-centric style. We, however, aim at enhancing the framework by adopting dumb gateway nodes besides having layered fog that will enhance the performance, reduce the latency, and provides more composition flexibility and options. Before explaining our proposed architecture in details, we explain the same scenario used in [10] with some changes that show the need for our suggested modifications along with a top level description of our approach.

3.1   Top Level Description

The environment specialist (end user) requires observations of many sensors (perhaps of same type or different types of sensors) as well as a the ability to trigger/actuate on actuator(s). To fulfil his request, we may have to hire those devices from different owners (same type of device may belong to different owners or come under different networks/gateways) or an average of observation values is required for some location which require readings from multiple sensors.

The Fog Node Manager in the cloud is in charge of selecting the fog node(s) and the gateway(s) associated with the requited Sensor(s)/Actuator(s) which may be in different locations. Figure 1 shows the network layers and Fig. 2 shows the top level diagram of the architecture along with the connections among layers.

The following subsections present a bottom-up layerwise illustrating of the architecture Followed by general explanation regarding reducing response time and failure plan taken care in the proposed architecture.

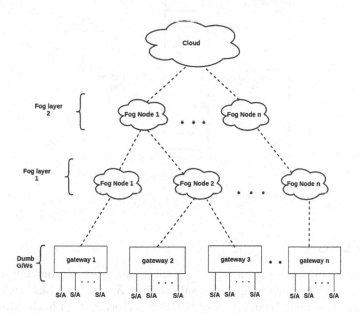

Fig. 1. Network layers of the proposed architecture

3.2   Things and Gateway Layer

As in depicted in Fig. 3 that shows the gateway architecture, Sensors and actuators are connected to dumb gateways- does not do any processing. The Actuator Selector receives the commands from the fog node (via MQTT message translator that makes it clear it is an actuating command and also extract the target actuator from the command) and applies the force on the targeted actuator. Similarly, Sensors and Observation Selector is responsible for sending each sensors observation values to the fog node via dedicated MQTT channels. The board pins are to connect the board with the physical sensors and actuators for observation receiving and applying commands respectively. The reason of having MQTT translator here is that MQTT does not have any fixed message forms hence the translator will take care of avoiding any clashes in the sensors observation, their channels, sensors IDs, time-tamp etc. the same which are agreed with the fog node.

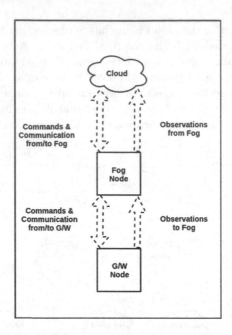

Fig. 2. Top level diagram

3.3 Fog Layer

Each fog node is responsible for managing the gateway(s), sensors and actuators assigned to it by the cloud assignment submodule. Among its tasks are; receiving observation, converting them into MongoDB and storing them in the observation data base. Figure 4 depicts Fog Node architecture. End users may have some cases with predetermined reaction to be taken, hence the monitoring and the decision will be done at the Specified Cases Manager in the fog node in witch the sensors and actuators associated with the case are assigned to. For example (in Fig. 1), an end user has rented sensors and actuators belongs to gateway 1, then the specified case monitoring will be assigned to fog node 1 in fog layer 1. If the sensors and actuators of end user interest are in gateway 2 or gateway 3, or some in both gateways, then the monitoring must be in fog node 2 in fog layer 1. But assume the sensors and actuators of end user interest belong to gateway 1 and gateway 3, then the monitoring must be in fog node 1 in fog layer 2. This monitoring plan holds while going up in the topology; in case there is no any more layer between fog layer 2 and the cloud, then cloud itself will monitor cases of sensors and actuator belong to gateways 1 and n.

Once Cloud assigns gateways, and sensors and actuators (G/W and S/A for short, respectively) to any fog node, their information is stored in the G/W and S/A database through the Fog Node Manager which is responsible for communication with the Cloud. The dashed line around G/W Management Agent and MQTT Translator in the fog node in Fig. 4 shows the logical grouping that is to

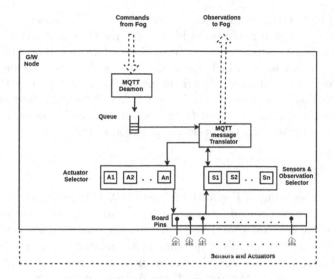

Fig. 3. Gateway architecture

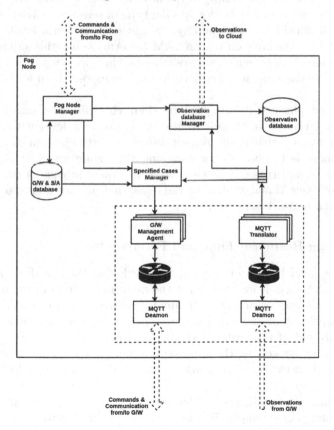

Fig. 4. FogNode architecture

be reassigned to the nearest fog node in case of failure of current fog node managing the gateway which is the main idea of failure plan that will be discussed in later subsection. Observations sent from G/W(s) will be received by MQTT translator to the queue in which the observations the Observation Database Manager is responsible of converting them into MongoDB and store them in the observation database. Whereas G/W Manager Agent receives the commands of applying force on a target actuator from Specified Cases Manager and sends it to the gateway in which the target actuator is connected.

3.4   Cloud Layer

In cloud side, as showing in Fig. 5, the Cloud G/W is the gateway between end users and the service provider. It receives the end user requests and delivers the virtual set of devices. Cloud G/W is also the gateway for devices owner to register their devices details. Core management send the request details to the Physical S/A Selection module which in turn is responsible for selecting the optimal devices as per the Service Level Agreement and previous customers review. We also take the availability of the device into consideration for selection; for example the end user request for specific type of sensor (i.e. from 2AM to 11 AM) cant be fulfilled by hiring one physical device as it is not available for the required duration (available only from 2AM to 5Am) hence this submodule will select one or more devices (which are available for the remaining time required) of the same type at the same required location and compose them for the required duration.

The selected physical devices will be then virtualized to create a virtual set for the end user, here the virtualization is on data level with respect to sensors observation which is the responsibility of Virtualization Manager. Fog G/W assignment is responsible for assigning the management of G/W to the nearest fog node and reassignment to any other node in case of fog node fail. The Specified Cases Manager monitor the cases that are not assigned to any fog node as per the network topology.

3.5   Reducing Response Time and Failure Plan

The use of fog layer between the cloud and deploying the specified cases monitoring in fog nodes will leads to starting the filtration process of the observation and applying the force on actuators (if any is required as per the specified cases) faster than having it at the cloud side irrespective to the filtration algorithms used which are out of the scoop of this work. This is the main way we reduce the response time for getting the end user's actuating command done faster for the prespecified cases; in other words we move the decision to lower level of architecture.

Failure Plan: The main goal of failure plan is to cope with any fog node failure situation. For example, in Fig. 1, if Fog Node 2 fails, then Fog G/W assignment manager in the cloud will reassign the management of gateway 2 to

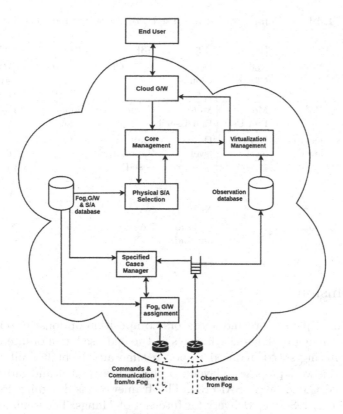

Fig. 5. Cloud-side architecture

Fog Node 1. That explains the dashed line around G/W Management Agent and MQTT translator in fog nodes in Fig. 4 that shows the units that are supposed to be migrated to the new management fog node

4   Comparison with Related Work

In this section we present a comparative evaluation of our proposed work against the aforementioned related work with respect to points listed in Table 1. That is we wish to emphasise the architectural aspects.

Regarding processing in [9] (*), the work did not state any thing about data flow or processing. Also with respect to failure plan (**), works [9, 11–13] do not adopt any concept like Fog or Edge nodes which is where and what we mean by failure plan in this point of comparison. This comparison is based on broad points (listed in the table) which highlight the main differences among works involved in the comparison; not in the performance of each work which can not be judged fairly without simulating all works under same situations.

Table 1. Comparison of proposed work with related work

Reference	[9]	[11]	[12]	[13]	[10]	SALFSD
Technology	Node-Cloud	Node-Cloud	Device-Cloud	Node-Cloud	Cloud Edge-Centric	G/W-Layered Fog-Cloud
Processing at	No*	Mote, End User	Mote, End User	Cloud	Edge, Cloud	Fog, Cloud
Virtualization	Device-level	Device-level	Device-level	Cloud (OpenStack based)	At Cloud	Cloud (Data-Centric, Virtual Actuators)
Failure plan	No**	No**	No**	No**	No	Yes
Working model	No	No	Only for android	No	No	No

5 Conclusion

To provide multiple sensors and actuators composition options (of sensors and actuators belonging to different gateways of the end used area of interest), with high data filtering speed, reduced response time and in built fault resistance plans, we in this work proposed a layered architecture (Things and gateway layer, Fog layer, and Cloud layer) SALFSD. The framework is designing for Sensing and Actuating as a Service Delivery for Internet of Things. The evaluation of the proposed framework will be performed in the future work by using real sensors and actuators devices in simulation environment or using iFogSim.

References

1. Al-Fuqaha, A., Guizani, M., Mohammadi, M., Aledhari, M., Ayyash, M.: Internet of Things: a survey on enabling technologies, protocols, and applications. In: 17th IEEE Communications Surveys Tutorials, pp. 2347–2376 (2015). https://doi.org/10.1109/COMST.2015.2444095
2. Botta, A., de Donato, W., Persico, V., Pescapé, V.: Integration of cloud computing and internet of things: a survey. In: 65th Future Generation Computer Systems, pp. 684–700 (2016). https://doi.org/10.1016/j.future.2015.09.021
3. Rao, B., Saluia, B., Sharma, N., Mittal, N., Sharma, S.: Cloud computing for Internet of Things & sensing based applications. In: Sixth International Conference on Sensing Technology (ICST), pp. 374–380, December 2012. https://doi.org/10.1109/ICSensT.2012.6461705
4. Rimal, B., Choi, E., Lumb, I.: A taxonomy and survey of cloud computing systems. In: 2009 Fifth International Joint Conference on INC, IMS and IDC, pp. 44–51, August 2009. https://doi.org/10.1109/NCM.2009.218
5. Gubbi, J., Buyya, R., Marusic, S., Palaniswami, M.: Internet of Things (IoT): a vision, architectural elements, and future directions. CoRR (2012)

6. Tordera, E., et al.: What is a fog node a tutorial on current concepts towards a common definition. CoRR (2016)
7. Al-khafajiy, M., Baker, T., Waraich, T., Al-Jumeily, D., Hussain, A.: IoT-Fog optimal workload via fog offloading. In: IEEE/ACM International Conference on Utility and Cloud Computing Companion (UCC Companion), pp. 359–364, December 2018. https://doi.org/10.1109/UCC-Companion.2018.00081
8. Al-Perera, C., Zaslavsky, A., Christen, P., Georgakopoulos, D.: Sensing as a service model for smart cities supported by Internet of Things. CoRR (2013)
9. Distefano, S., Merlino, G., Puliafito, A.: Sensing and actuation as a service: a new development for clouds. In: IEEE 11th International Symposium on Network Computing and Applications, pp. 272–275, August 2012. https://doi.org/10.1109/NCA.2012.38
10. Satpathy, S., Sahoo, S., Turuk, A.: Sensing and actuation as a service delivery model in cloud edge centric internet of things. Future Gener. Comput. Syst. (2018.) https://doi.org/10.1016/j.future.2018.04.015
11. Distefano, S., Merlino, G., Puliafito, A., Vecchio, A.: A hypervisor for infrastructure-enabled sensing clouds. In: 2013 IEEE International Conference on Communications Workshops (ICC), pp. 1362–1366, June 2013. https://doi.org/10.1109/ICCW.2013.6649449
12. Distefano, S., Merlino, G., Puliafito, A.: A utility paradigm for IoT: the sensing cloud. Pervasive Mobile Comput. 127–144 (2015). https://doi.org/10.1109/ICCW.2013.6649449
13. Longo, F., Bruneo, D., Distefano, S., Merlino, G., Puliafito, A.: Stack4Things: a sensing-and-actuation-as-a-service framework for IoT and cloud integration. Annales des Télécommunications, 53–70 (2017)
14. OpenStack. https://docs.openstack.org/stein/

Smartphone Based Outdoor Navigation and Obstacle Avoidance System for the Visually Impaired

Qiaoyu Chen[1], Lijun Wu[1]($\boxtimes$) (iD), Zhicong Chen[1](iD), Peijie Lin[1], Shuying Cheng[1], and Zhenhui Wu[2]

[1] College of Physics and Information Engineering, Fuzhou University, Fuzhou 350116, China
lijun.wu@fzu.edu.cn
[2] State Grid Fuzhou Electric Power Supply Company, Fuzhou 350116, China

Abstract. Interlaced roads and unexpected obstacles restrict the blind from traveling. Existing outdoor blind auxiliary systems are bulky or costly, and some of them cannot even feedback the type or distance of obstacles. It is important for auxiliary blind systems to provide navigation, obstacle detection and ranging functions with affordable price and portable size. This paper presents an outdoor navigation system based on smartphone for the visually impaired, which can also help them avoid multi-type dangerous obstacles. Geographic information obtained from GPS receiving module is processed by professional navigation API to provide directional guidance. In order to help the visually impaired avoid obstacle, SSD-MobileNetV2 is retrained by a self-collected dataset with 4500 images, for better detecting the typical obstacles on the road, i.e. car, motorcycle, electric bicycle, bicycle, and pedestrian. Then, a lightweight monocular ranging method is employed to estimate the obstacle's distance. Based on category and distance, the risk level of obstacle is evaluated, which is timely conveyed to the blind via different tunes. Field tests show that the retrained SSD-MobileNetV2 model can detect obstacles with considerable precision, and the vision-based ranging method can effectively estimate distance.

Keywords: Blind auxiliary system · Walking navigation · Ranging · SSD-MobileNetV2 · Smartphone

1 Introduction

According to the reports from the China Disabled Persons' Federation, there are more than 12.63 million visually impaired people in China [1]. The visually impaired are unable to travel independently since they are easy to lose their directions, not mention that the complex traffic environment poses a great threat to their safety. Therefore, there is an urgent need for technologies that can help the blind find their way and avoid obstacles.

© Springer Nature Switzerland AG 2019
R. Chamchong and K. W. Wong (Eds.): MIWAI 2019, LNAI 11909, pp. 26–37, 2019.
https://doi.org/10.1007/978-3-030-33709-4_3

There are already various auxiliary equipments available for the blind, including white cane, guide dog and some electronic aids such as AngelEye [2] and Brian Port [3]. As an economic tool, white cane is universally used by the blind. However, it can only sense the obstacle within a limited range, and the users are easy to be worn out since they have to concentrate themselves all the time. Guide dog is a good solution for helping the blind travel, but they are very scarce and difficult to be domesticated. Both AngelEye and Brian Port are wearable auxiliary blind glasses. AngelEye provides navigation, object recognition, color recognition and illumination recognition functions to users. Brian Port reconstructs images in the brain through specially designed electrodes. These two products are powerful, but their prices range from $1300 to $7300 [2,3], which is too expensive for most visually impaired people.

Recently, the smartphone-based navigation auxiliary system for the blind has attracted much attention. Several navigation designs for visually impaired people are presented in [4–6]. Article [4] presents a campus navigation system for visually impaired people named Divya-Dristi with a 3-tier architecture. Tier-1 includes a smartphone and a sonar device. Smartphone is used to obtain GPS data by the built-in GPS reviving module, and sonar device is employed to detect obstacles on the road. Tier-2 is a cloud server that queries the Tier-3 to extract the navigation details. Tier-3 is a geographic database, which can be updated according to the user's location. In Divya-Dristi, the sonar device is unable to distinguish the types of obstacles. Besides that, the Divya-Dristi is not integrated enough since the sonar and the smartphone are separated. In [5], Nawin somyat et al. designed a navigation app that can help the blind travel locally in Thailand, which is named NavTu. Usually, NavTu gets navigation services from Google Maps [7]. Moreover, it can store custom paths that are not provided on Google Maps. NavTu uses a method based on image vanishing point to range the stationary obstacles, such as the concrete electric poles or the walls. A guidance system for the blind proposed in [6] obtains navigation services from OpenStreetMap [8], which conveys the direction information to the blind by a vibrating sole. However, this system still relies on white cane or guide dog to avoid obstacles. To summarize, the existing systems are either composed of several parts with low integration or not able to provide the obstacles classification and ranging. Therefore, in this work, we propose a solely smartphone based auxiliary system, which can help the blind find their way and avoid obstacles depend on category and distance.

2 Related Work

Recently, the idea of deep learning has been widely applied in the field of object detection. The utilization of deep learning makes significant progress in the detection precision. Generally speaking, there are two main branches for object detection algorithms: the proposal-based branch and the proposal-free branch. Basically, the proposal-based algorithms extract the regions that are most likely to contain object by selective search [9] at first, then analyze these regions with convolutional neural network (CNN), after which the areas that actually contain objects are picked out and the class and the fine-tuning coordinates are determined as well. R-CNN [10], SPP-net [11], Fast R-CNN [12] and Faster R-CNN [13] are classical proposal-based algorithms. Although above algorithms can achieve an excellent precision, the process of region proposal inevitably slows down the speed. For example, the Faster R-CNN can run at 7 frames per second (FPS), which is too slow to meet the requirements of real-time tasks. The proposal-free algorithms abandon generating the region proposals, which remarkably improves the detection speed. You only look once (YOLO) [14] series and single shot multibox detector (SSD) [15] are excellent proposal-free algorithms. YOLO regards object detection as a single regression problem, and directly outputs the bounding box coordinates and class probabilities by a single CNN. YOLO makes amazing breakthroughs in speed, the regular YOLO can reach 45 FPS, while the simplified Fast YOLO can even reach 155 FPS. YOLOv2 [16] adds the passthrough layer and concatenates the 13×13 and 26×26 feature maps to extract fine-grained information, which effectively reduces the miss detection rate of small objects. Then, Joseph Redmon et al. renew YOLOv2 and propose YOLOv3 [17]. YOLOv3 combines the feature maps of three scales for prediction, so that the recall rate is improved once again. SSD is another important proposal-based algorithm. The main idea of SSD is to use convolutional filters to predict bounding box on multi-scale feature maps and initialize a better bounding box sizes by anchors. The detection speed of SSD is faster than YOLOv1, and the detection precision is comparable to Faster R-CNN.

Above brilliant object detection algorithms need to rely on the backbones to extract features, such as VGGnet [18], Darknet-19 [16], Darknet-53 [17] and so on. However, these networks are too large to be deployed on platforms with limited computation. In 2017, Google team proposed a lightweight deep neural network MobileNetV1 [19] for smartphone and embedded device. And in 2018, they proposed a new version MobileNetV2 [20]. Several experiments in [19, 20] show that the combination of SSD algorithm and MobileNet can achieve considerable accuracy while run smoothly on smartphone. Therefore, this work combines SSD and MobileNetV2 for obstacle detection.

3   Walking Navigation

In this section, the navigation module is introduced, including location, voice entry and voice navigation.

3.1   Location

The smartphone built-in GPS receiving module receives GPS data from satellites, which are transmitted over the Internet to Baidu Maps server. The server will return the corresponding latitude, longitude, altitude, location description, etc. to the smartphone. Baidu Maps server and smartphone can communicate via 4G or the upcoming 5G. Upon the location is completed, smartphone will broadcast the current location to the blind.

3.2   Voice Entry

It is difficult for the blind to manually text the destination at a specified position on the screen. For this purpose, we realize voice-based destination entry method for the blind. In detail, a voice will guide the blind to press the screen and begin to speak out the destination. Then, the smartphone converts the audio into a text address that is parsed into longitude and latitude.

3.3   Voice Navigation

The voice navigation service is provided by the Baidu Maps. From the previous location step and voice entry step, we can obtain the latitude and longitude of the starting point and the destination, which will be transmitted to the Baidu Maps server for returning the detailed navigation information to the smartphone. Walking guidance are broadcast by voice during the whole travel, and the users will hear professional navigation information such as "go straight for 500 m", "turn left after 50 m", etc.

4   Obstacle Avoidance

In this section, we will describe the components of the obstacle avoidance, including obstacle detection, ranging, and obstacle warning. Firstly, we introduce the obstacle detection model SSD-MobileNetV2. Secondly, we explain the principle and implementation of ranging. At last, the obstacle warning step is briefly described.

4.1   Obstacle Detection

In this paper, the SSD-MobileNetV2 model is applied to detect obstacles. MobileNetV2 extracts the features from the original image, and SSD predicts the obstacles based on the feature maps.

In MobileNetV2, the convolutional layer is an important structure for extracting image features. Generally, the deeper convolutional layers, the better features. However, the deepening of the network will cause calculation explosion, gradient disappearance or the gradient explosion, which makes deep learning models difficult to deploy on platforms with limited computing power and memory. MobileNetV2 replaces standard convolution with depthwise separable convolution, a $D_K \times D_K \times M$ standard convolutional filter is divided into M $D_K \times D_K \times 1$ depthwise convolutional filters and a $1 \times 1 \times M$ pointwise convolutional filter. With the improvement of convolution structure, the convolution computational complexity is reduced by 8 times to 9 times [19]. Moreover, MobileNetV2 proposes an inverted residual block that can effectively prevent gradient disappearance or gradient explosion. Inverted residual block first uses 1×1 convolution with ReLU6 [21] to increase channels, then performs a 3×3 depthwise separable convolution operation with ReLU6, finally restores the original channels with 1×1 convolution and adds the input with output [20].

After the basic features generated by MobileNetV2, several layers of depthwise separable convolution are operated to generate several feature maps with decreasing scales. SSD performs on multi-scale feature maps to predict multi-scale objects. At prediction time, each feature map will be evenly divided into cells, every cell predicts k bounding boxes and c category confidences. For each category, top n bounding boxes will be retained. Then the non maximum suppression (NMS) is performed to filter out the bounding boxes with large overlap, and the rest is detection results. In this paper, we use mAP to judge the effectiveness of obstacle detection model, mAP reflects the mean of all average precision (AP) of each category, where the precision is defined in (1):

$$Precision = \frac{True\ Positive}{True\ Positive + False\ Positive}.\qquad(1)$$

4.2   Obstacle Ranging

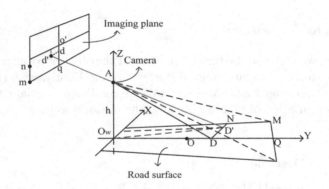

Fig. 1. Illustration of monocular ranging.

In this part, we introduce a monocular vision-based method for ranging road obstacles. The principle of monocular ranging is illustrated in Fig. 1, the O, D, D', Q, M and N on the road surface correspond to o' (imaging plane center), d, d', q, m and n on the imaging plane respectively, Q is farthest measurable points. In this model, the camera is fixed on the blind and tilted to the road. Taking distance $O_w D'$ as an example to illustrate the ranging method. Distance $O_w D'$ can be calculated according to the Pythagorean theorem with known vertical distance $O_w D$ and horizontal distance DD'. The calculation methods of $O_w D$ and DD' are detailed separately in below.

Vertical distance $O_w D$ calculation method: In order to make the principle of $O_w D$ calculation clearly, the Fig. 1 is mapped to the Y-axis as shown in Fig. 2. Let the line BO be perpendicular to the line AO, and from geometric knowledge we can get:

$$\triangle AOB \sim \triangle Ao'q,\qquad(2)$$

$$\triangle AOC \sim \triangle Ao'd. \tag{3}$$

Therefore,

$$\frac{OC}{OB} = \frac{AO \times \tan g}{AO \times \tan \beta} = \frac{\tan g}{\tan \beta} = \frac{\tan(|\theta - \alpha|)}{\tan \beta} = \frac{o'd}{o'q}, \tag{4}$$

that is,

$$\frac{\tan(|\theta - \alpha|)}{\tan \beta} = \frac{o'd}{o'q}, \tag{5}$$

so

$$\alpha = \theta \pm \arctan(\frac{o'd \times \tan \beta}{o'q}). \tag{6}$$

Since

$$\tan \alpha = \frac{h}{O_w D}, \tag{7}$$

thus

$$O_w D = \frac{h}{\tan \alpha} = \frac{h}{\tan(\theta \pm \arctan(\frac{o'd \times \tan \beta}{o'q}))}. \tag{8}$$

where h is the height of the camera from the ground, β is camera vertical half-field angle, and $o'q$ is half of the image height, these are all constant. β can be obtained by (9):

$$\beta = \arctan(\frac{H \times l}{2 \times f}). \tag{9}$$

in which H is the image height, l is the pixel size, and f is the focal length. Besides that, in formula (8), θ is camera pitch angle, $o'd$ is the vertical length from the point d' to the image center point o', they are both variables. As shown in the Fig. 3, θ and δ are complementary, so θ can be figured out by (10):

$$\theta = 90° - \delta, \tag{10}$$

where δ can be measured from the built-in direction sensor in the smartphone. Moreover, $o'd$ can be calculated according to detection box bottom-middle coordinate (x, y). Taking Fig. 4 as an example, (x, y) can be figured out from (11) and (12):

$$x = \frac{x_{min} + x_{max}}{2}, \tag{11}$$

$$y = y_{max}, \tag{12}$$

then the $o'd$ can be derived from (13):

$$o'd = |y_{max} - o'q|. \tag{13}$$

Horizontal distance DD' calculation method: As shown in the Fig. 5, the plane ADN in Fig. 1 is extracted to calculate DD'. Derived from geometric knowledge we can know:

$$\triangle ADD' \sim \triangle Add', \tag{14}$$

$$\triangle ADN \sim \triangle Adn. \tag{15}$$

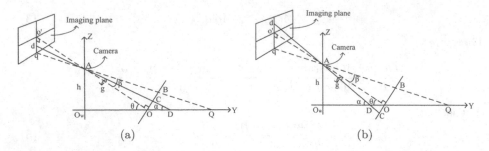

Fig. 2. Calculation illustration of $O_w D$: (a) D is farther than O, (b) D is closer than O.

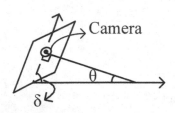

Fig. 3. Illustration of camera pitch angle.

Fig. 4. Illustration of (x, y).

Accordingly,

$$\frac{DD'}{DN} = \frac{DD'}{h' \times \tan\gamma} = \frac{dd'}{dn}, \tag{16}$$

that is,

$$DD' = \frac{dd'}{dn} \times h' \times \tan\gamma, \tag{17}$$

cause

$$h' = \sqrt{h^2 + O_w D^2}, \tag{18}$$

hence,

$$DD' = \frac{dd'}{dn} \times \sqrt{h^2 + O_w D^2} \times \tan\gamma, \tag{19}$$

where γ is the horizontal camera half-field angle, dn is half the width of the image, they are all constant. Among them, γ can be obtained by (20):

$$\gamma = \arctan(\frac{W \times l}{2 \times f}), \tag{20}$$

in which W is the image width. Besides that, in formula (19), dd' is the horizontal width from the point d' to the image center point o', which can be derived from the (21):

$$dd' = |x - dn|, \tag{21}$$

Finally, according to the Pythagorean theorem, the distance between the object and the camera can be figured out from (22):

$$dis = \sqrt{DD'^2 + O_w D^2}.\tag{22}$$

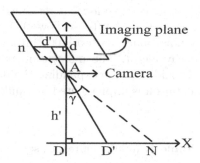

Fig. 5. Calculation illustration of DD'.

4.3   Obstacle Warning

Seven tunes are used to indicate the risks of obstacles to the blind. The total risk of an obstacle, is defined by the sum of its distance risk factor and its category risk factor. The distance is divided from 0 m to 10 m into three intervals, i.e. $dis\_intervals = \{(0 \text{ m}, 5 \text{ m}], (5 \text{ m}, 7 \text{ m}], (7 \text{ m}, 10 \text{ m}]\}$, and the corresponding risk factors of distances are marked as $dis\_risks = \{3, 2, 1\}$. Normally, the main obstacles on the road are pedestrians, bicycles, motorcycles, and cars. In China, there is also one of the most common vehicles—electric bicycles. Therefore, the risk factors of different categories are marked as $category\_risks = \{1, 2, 3, 4, 5\}$, which corresponds to the risk factor of pedestrian, bicycle, electric bicycle, motorcycle and car respectively. Therefore, seven total risk levels can be obtained by sum the distance risk factor and the category risk factor, i.e. $risk\_levels = \{2, 3, 4, 5, 6, 7, 8\}$. The larger the number, the higher the risk. Seven tunes are correspond to seven risk levels from low to high. If there are multiple obstacles in an image, the app only indicates the three most dangerous ones, so as to avoid the confusion caused by too many broadcasts.

5   Experiments and Discussions

In order to retrain the obstacle detection model SSD-MobileNetV2, we collect pictures of five categories, including cars, motorcycles, electric bicycles, and pedestrians in local traffic environment to form a dataset containing about 4500 images. The sample number of each category is comparable. The self-collected dataset is randomly divided into training set and testing set in a ratio of 3:1, in which the training set contains about 3375 images and the testing set contains about 1125 images. The dataset is done when the obstacles are moving or stationary. We annotate the dataset with LableImg, and convert the annotations and images to tfrecord format for being used in the Tensorflow framework. The car, motorcycle, electric bicycle, and pedestrian are respectively annotated as "car", "motorbike", "elebike", "bicycle", and "person". All images are resized to 300×300 for training the model.

SSD-MobileNetV2 object detection model is retrained for about 180k steps on the self-collected training dataset by Tensorflow Object Detection API. We use exponential decay learning policy with a 0.009 initial learning rate, a 1000 decay steps and a 0.95 decay factor. The momentum is 0.9 with a 0.9 decay, and the batch size is 48. In addition, for expanding the dataset and preventing overfitting, dataset augmentation is implemented by random horizontal flip and random crop.

Table 1. Results on testing set.

Model	mAP	person	bicycle	elebike	motorbike	car
SSD-MobileNetV2	84.6%	69.0%	86.6%	90.9%	80.2%	96.2%

Fig. 6. Detection examples on testing set.

We apply the retrained model to the testing set to evaluate performance. As shown in Table 1, we finally got a 84.6% mAP on the testing set that proves the retrained model can effectively learn the representation of various obstacles. The AP of "person" is lower than that of other obstacles, which is due to the variability of body posture. We plan to enrich the training set of "person" to improve its AP. Photos of some testing results are shown in Fig. 6. The retrained can detect the mentioned five types of obstacles in this paper, but there are still some small objects are missed. The reasons for missing detection can be divided into two aspects: On one hand, small objects occupy fewer pixels and contain less obvious features in images, which weakens the responses of convolution. With the superposition of convolutional operations, the useful information of small objects may disappear at the end of network. On the other hand, feature maps with different levels can provide favorable detection clues to each other. However, in SSD, the detection operation is performed independently on each feature map, which loses many important contextual information for detection.

Table 2. Ranging results.

Real values (m)	1.00	2.00	3.00	4.00	5.00	6.00	7.00	8.00	9.00	10.00
Measured values (m)	1.06	2.10	3.06	4.04	5.02	6.09	7.02	8.03	9.09	10.02
Errors (m)	0.06	0.10	0.06	0.04	0.02	0.09	0.02	0.03	0.09	0.02

Experiments are also carried out to test the ranging methodology in local road environments. We fix the smartphone on a person and keep the height of the camera is 1.20 m above the ground. We regard a car as an obstacle and measure the distance every 1 m in the range from 0 m to 10 m. The distance measurement results are shown in Table 2. As we can see, the measured distances are close to the real distances at the meter level. The deviations of camera height and detection box cause slight ranging errors. Usually, people estimate the obstacle is about "7 m ahead" instead of "7.654 m ahead". Therefore, the ranging method is applicable for blind auxiliary system.

| (a) | (b) | (c) |

Fig. 7. Results of detection and ranging on local roads: (a) The real distances of car and bicycle are 4.30 m and 3.20 m, (b) The real distances of bicycle and motorbike are 2.10 m and 2.80 m, and (c) The real distances of bicycle and elebike are 2.50 m and 1.35 m.

We also integrate the navigation system, the obstacle detection algorithm and the ranging method on a smartphone with Android platform, and then test the obstacle detection and the ranging method on the local roads. The bounding boxes with the category label, category confidence and ranging result are displayed on the phone's screen. Several experimental results are shown in Fig. 7. As one can see, all the obstacles in Fig. 7 can be correctly detected, and the results of ranging are relatively close to the real distance. To sum up, the detection model and ranging method of this work are effective and feasible.

6   Conclusion

In this paper, we implement a smartphone based blind auxiliary system that can provide GPS navigation, obstacle detection, ranging, and musical tune based obstacle risk warning with a friendly voice interaction mode. In addition, we collect an object detection dataset with 4500 images including cars, pedestrians, bicycles, electric bicycles, and retrain the object detection model SSD-MobileNetV2. Preliminary experiments results show that the retrained model affords a good support for obstacle detection with considerable accuracy, but the detection results of small objects is not satisfactory. The vision-based ranging method can effectively estimate the distance of road obstacles at meter level. In the future, we plan to improve the ranging method robustness to different heights, and further promote the detection mAP, while improving detection performance of small objects.

Acknowledgement. This work is financially supported in parts by the Fujian Provincial Department of Science and Technology of China (Grant No. 2019H0006 and 2018J01774), the National Natural Science Foundation of China (Grant No. 61601127), and the Foundation of Fujian Provincial Department of Industry and Information Technology of China (Grant No. 82318075).

References

1. Brief Data of the National Basic Information Database of Persons with Disabilities. http://www.cdpf.org.cn/sjzx/cjrgk/201206/t20120626_387581.shtml. Accessed 6 July 2019
2. AngleEye. http://www.nextvpu.com/cn/ProInfo.aspx?proid=1&categoryid=6. Accessed 6 July 2019
3. Brian Port. https://www.wicab.com/. Accessed 6 July 2019
4. Dutta, S., Barik, M., Chowdhury, C., Gupta, D.: Divya-Dristi: a smartphone based campus navigation system for the visually impaired. In: International Conference on Emerging Applications of Information Technology, pp. 1–3. IEEE (2018)
5. Somyat, N., Wongsansukjaroen, T., Longjaroen, W., Nakariyakul, S.: NavTU: android navigation app for Thai people with visual impairments. In: International Conference on Knowledge and Smart Technology, pp. 134–138. IEEE (2018)
6. Velázquez, R., Rodrigo, P.: An outdoor navigation system for blind pedestrians using GPS and tactile-foot feedback. Appl. Sci. 8(4), 578 (2018)

7. McMahon, D.: Effects of digital navigation aids on adults with intellectual disabilities: comparison of paper map, Google maps, and augmented reality. J. Spec. Educ. Technol. **30**(3), 157–165 (2015)

8. Haklay, M., Weber, P.: OpenStreetMap: user-generated street maps. IEEE Pervasive Comput. **7**(4), 12–18 (2008)

9. Uijlings, J.: Selective search for object recognition. Int. J. Comput. Vis. **104**(2), 154–171 (2013)

10. Girshick, R., Donahue, J., Darrell, T., Malik, J.: Rich feature hierarchies for accurate object detection and semantic segmentation. In: CVPR, pp. 580–587. IEEE (2014)

11. He, K.: Spatial pyramid pooling in deep convolutional networks for visual recognition. IEEE Trans. Pattern Anal. Mach. Intell. **37**(9), 1904–1916 (2015)

12. Girshick, R.: Fast R-CNN. In: ICCV, pp. 1440–1448. IEEE (2015)

13. Ren, S.: Faster R-CNN: towards real-time object detection with region proposal networks. IEEE Trans. Pattern Anal. Mach. Intell. **39**(6), 1137–1149 (2015)

14. Redmon, J., Divvala, S., Girshick, R., Farhadi, A.: You only look once: unified, real-time object detection. In: CVPR, pp. 779–788. IEEE (2016)

15. Liu, W., et al.: SSD: single shot MultiBox detector. In: Leibe, B., Matas, J., Sebe, N., Welling, M. (eds.) ECCV 2016. LNCS, vol. 9905, pp. 21–37. Springer, Cham (2016). https://doi.org/10.1007/978-3-319-46448-0_2

16. Redmon, J., Farhadi, A.: YOLO9000: better, faster, stronger. In: CVPR, pp. 6517–6525. IEEE (2017)

17. Redmon, J., Farhadi, A.: YOLOv3: an Incremental Improvement. arXiv:1804.02767 (2018)

18. Simonyan, K., Zisserman, A.: Very deep convolutional networks for large-scale image recognition. arXiv:1409.1556 (2015)

19. Howard, A., et al.: MobileNets: efficient convolutional neural networks for mobile vision applications. arXiv:1704.04861 (2017)

20. Sandler, M., Howard, A., Zhu, M., Zhmoginov, A., Chen, L.: MobileNetV2: inverted residuals and linear bottlenecks. In: CVPR, pp. 4510–4520. IEEE (2018)

21. Krizhevsky, A., Hinton, G.: Convolutional deep belief networks on CIFAR-10. Unpublished manuscript 40(7), 1–9 (2010)

Recent Developments in Recommender Systems

Jia-Ming Low[1,2](✉), Ian K. T. Tan[3]⬢, and Choo-Yee Ting[1]⬢

[1] Multimedia University, 63100 Cyberjaya, Selangor, Malaysia
[2] Priority Dynamics Sdn Bhd, Garden Shoppe @ One City, 47650 Subang Jaya, Selangor, Malaysia
`lenox@prioritydynamics.com`, `cyting@staff.mmu.edu.my`
[3] Monash University Malaysia, Bandar Sunway, 47500 Subang Jaya, Selangor, Malaysia
`ian.tan1@monash.edu`
`http://www.mmu.edu.my`
`http://www.monash.edu.my/IT`

Abstract. With greater penetration of online services, the use of recommender systems to predict users' propensity for continuous engagement becomes crucial in ensuring maximum revenue. There are many challenges, such as the cold start problem and data sparsity, that are continuously being addressed by a myriad of techniques in recommender systems. This paper provides insights into the trends of the techniques used for recommender systems and the challenges they address. With the insights; deep learning, matrix factorization or a combination of both can be used in addressing the data sparsity challenge.

1 Introduction

Recommender systems (RS) refer to computerized systems with the purpose of filtering out potentially items that are of little or no interest to the users. The output of the systems will be a smaller set of items that are of interest to the users. It is a form of information filtering, similar to how news editors aggregate news and present the most relevant news only. With the advent of the Internet, information overload is a challenge and recommender systems become crucial to ensure only relevant items are presented and other items are filtered out.

One of the earliest known use of filtering techniques as a recommender system was a system called Tapestry. Tapestry allows users to search for documents based on keywords and also search based the reactions marked by other users. From then on, RS have been improved and applied for various systems; such as the application of RS by Netflix to improve users' experience for their video streaming services. Netflix provided recommendations by using the users' past activities such as *watched*, *liked*, and *favourite videos*. The popular e-commerce website, Amazon used RS to increase their sales. When a user purchases an item, the system will recommend other products purchased by other users according

© Springer Nature Switzerland AG 2019
R. Chamchong and K. W. Wong (Eds.): MIWAI 2019, LNAI 11909, pp. 38–51, 2019.
https://doi.org/10.1007/978-3-030-33709-4_4

to the purchased item. Amazon patented this approach, the collaborative recommendations using item to item similarity mappings.

In this paper, we provide an overview of the different types of approaches to RS and the various main challenges of implementing RS. This will be followed by recent developments in the approaches to the challenges, mainly for the last 4 years.

The objective of this paper is to provide a concise brief to other researchers who would like to jump start and conduct research in this established area. This paper ends with some suggestions on future work in the area of recommender systems.

2 Types of Recommender Systems

RS can be categorized into two groups, personalized and non-personalized. Personalized RS will generate recommendations based on the users' information. User will get different recommendations according to their own historical behaviour. Non-personalized RS will generate recommendation based on the averaging of other users' behaviour. The recommendation does not depend on the current user. Based on the information used, RS can be classified into few approaches [13].

2.1 Content Based RS

Content based approach uses textual information of items to generate recommendations. This approach assumes that users will like the similar item that they liked before. Content based approach is able to give recommendations without users' information. However, it is limited to the items' initial descriptions or features.

2.2 Collaborative Filtering RS

Collaborative Filtering (CF) approach makes the recommendations by using the users' ratings on the items. This approach assumes that users will be interested on the items that other similar users liked. Also, this approach does not require item features to generate recommendations. However, it suffers from cold start problem. The cold start problem is when the system cannot make recommendations for new users or recommend new items as lack of ratings on the users or items respectively.

2.3 Demographic Filtering RS

Demographic filtering approach utilizes the users' demographic information to make recommendations. The demographic information are age, gender, occupation, education background and others. This approach assumes that users belong to a certain demographic cluster will have similar interest or preferences. This approach can recommend items without knowing the ratings of the items or users rated. However, demographic information is difficult to obtain.

2.4   Hybrid RS

Each approach of RS has its pros and cons. Hybrid approach neutralizes the shortcomings of the approaches by combining them. The most popular hybrid recommender system is combining content-based approach and collaborative approach.

2.5   Context Aware RS

After the Netflix Prize competition[1], different forms of users input was attached great importance. The accuracy of RS was boosted by utilizing both implicit and explicit feedback of the users [16]. Context-aware RS utilizes not only users' ratings but also users' feedback and contextual information. It is one of the recent trending approaches.

3   Challenges of Recommender System

In some circumstances, ratings data is not available. However, the completeness of ratings data has direct impact on RS performance. Several research works have been proposed to alleviate the problem. There are three challenges that are commonly discussed [13]. The challenges are briefly elaborated in the next 3 subsections.

3.1   Cold Start

The cold start problem occurs when new users or new items appear in the system. So, cold start problem can be classified as user cold start and item cold start

User Cold Start. When a new user joins the RS, the system does not have enough information on user's purchase history or previous ratings of items. As such, the system can't make recommendation for the user.

Item Cold Start. When a new item added into the RS, system does not have enough ratings related to this item. So that, the new item will not be able to recommend to users.

3.2   Sparsity

For non cold start situations, recommender systems are dependent on historical data of the users or of the items. The input to the recommender system is stored in a matrix where one dimension represents the users and the other dimension represents the items. For example, in an online music streaming service that has a large number of users and items (music), each user would have listened to a set

---

[1] https://www.kaggle.com/netflix-inc/netflix-prize-data.

of songs and they may or may not have rated a small set of songs. This would result in a sparse matrix where it will have many empty cells. The sparse matrix will not be effective for most recommender system algorithms as there are too many unknowns and would create inaccurate recommendations.

3.3 Scalability

As the users and items increase, RS needs more resources to process the matrix and generate the recommendations accurately and quickly. However, scalability issue is rarely been discussed among the publications in recent years. The improvement of computing hardware and application of nearest neighbor algorithm, scalability is not a big issue anymore.

4 Literature Review

Among the techniques stated in Sect. 2, CF is highly accepted as a RS technique for its good prediction performance. CF can be categorized into two approaches; memory based and model based. Memory based CF typically makes use of the nearest neighbors' preferences to generate recommendations. Similarity measures is the commonly used technique in memory based CF. Model based CF typically makes use of a pre-learned users rating model to generate recommendations. Latent factor model is one of the main areas of model based CF. Different forms of users input were integrated into RS to get better performance. Many approaches were utilized to transform users' feedback to suit the RS environment.

In addition to CF, many other approaches have been proposed in recent years for RS. These approaches have been categorized as *similarity measures, association rules, segmentation, matrix or tensor factorization, social network, machine learning, deep learning, sentiment analysis* and *hybrid* approaches.

4.1 Similarity Measures Approach

Similarity measures play an important role, especially in neighborhood based CF. They are used for selecting the nearest neighbour and determine the importance level of the neighbour's rating in the prediction. They have significant impact on the prediction accuracy and performance. Mu [24] proposed an improved similarity measure called Common Pearson Correlation Coefficient with Hellinger Distance to enhance the versatility of similarity measure and to alleviate the sparsity problem. Wang *et al.* [39] proposed a music recommendation system by using a modified random walk algorithms.

4.2 Association Rules Approach

Associative Rules approach is a techniques to discover hidden relationship among items. Cold Start problems can be avoided by using this approach because it does not require complex rating system. Osadchiy *et al.* [27] proposed a recommender system based on pairwise association rules. Yi *et al.* [46] proposed a library personalized recommendation service method based on improved association rules that used the Artificial Bee Colony (ABC) algorithm. Convertini *et al.* [5] proposed a RS using hybrid fuzzy association to improve the accuracy.

4.3 Segmentation Approach

There are some RS using segmentation technique to categorized the users into clusters. This technique is used for selecting neighbours of the target user efficiently and alleviates the scalability problem. Banerjeee *et al.* [2] proposed a movie recommendation system using particle swarm optimization where users were categorized into clusters according to their demographic information, movie ratings and movie genres. Subramaniyaswamy *et al.* [33] proposed an adaptive K-Nearest Neighbour (KNN) based RS through mining of user preferences. Logesh *et al.* [20] had proposed bio-inspired swarm intelligent-based clustering approach to enhance the stability of RS. Logesh *et al.* [21] proposed an urban trip recommendation system that utilised quantum-behaved particle swarm optimization to enhance the performance of the system. Li *et al.* [19] proposed a music personalized RS based on improved KNN algorithm.

4.4 Machine Learning Approach

Machine learning techniques have been applied in RS. This approach allows RS to generate diverse recommendations. It is also able to learn the preference changes of users. Aghdam *et al.* [12] proposed a context-aware RS using Hierarchical Hidden Markov Model (HMM) where the model recognized the changes of users' preferences over-time, and is able to predicted next diverse context for users. Xu *et al.* [42] proposed a RS based on eXtreme Gradient Boosting Classifier that utilized behavior records of consumers and improved the system performance.

4.5 Matrix or Tensor Factorization Approach

Matrix factorization is one of the latent factor models. This algorithm became popular during the Netflix Prize Challenge. The base of this algorithm is characterizing both users and items into vectors of factors that represent the rating patterns. The algorithm was also allowed to include additional information. This flexibility is the space that can be discovered and researched further. Lee *et al.* [17] proposed a collaborative filtering RS by using uninteresting items that the value of items' uninterestingness of users were calculated and filled into the sparse rating matrix. Feng *et al.* [7] proposed a faster randomized singular value decomposition (SVD) that accelerated the matrix completion without loss of accuracy. Luo *et al.* [23] proposed a co-SVD model that utilized tags and time information to enrich the data source and alleviated over-fitting issues. Yang *et al.* [45]

proposed an item-diversity-based collaborative filtering algorithm that added variance of minimization regularization term into the factorization process in order to generate diversity recommendations. Symeonidis *et al.* [34] proposed courses recommendation system by using multi-model matrix factorization with side information. Tahmasbi *et al.* [36] proposed a movie recommendation system that exploited a proposed weighting schema to capture temporal dynamics of user preferences. Gouvert *et al.* [9] proposed a negative binomial matrix factorization for RS to analyze over-dispersed count data. Zheng *et al.* [52] proposed a unified probabilistic matrix factorization (PMF) recommendation algorithm that constructed three latent feature vectors from user-resource rating matrix, user-tag tagging matrix and resources-tag correlation matrix to improve the accuracy of RS. Wang *et at.* [38] proposed a personalized time-aware tag recommendation using tensor factorization that able to recognized users' temporal dynamic preferences and capture their potential interests.

4.6   Sentiment Analysis Approach

After Netflix Prize competition, it has been proven that the performance of RS was improved by utilizing different form of users' input. Generally, people will search for related information before they make decisions. Feedback from others influence their decision making. By applying sentiment analysis on users' reviews or comments, some insights for the related items can be obtained as extra features of RS. Accuracy of the system can be improve by the adding the extra features. Han *et al.* [11] proposed a trip RS that implemented Sentiment Utility Logistic Model (SULM). Wang *et al.* [37] proposed opinion rating aggregation methods for movie RS that applied sentiment classification techniques on movie reviews to determine the polarity of each reviews and then aggregated them as a scores for selecting top *n* movies. Preethi *et al.* [30] proposed using users' reviews to recommend places close to the users' current location. Nabil *et al.* [26] proposed an article RS that analyzed the consumers reviews of articles on twitters to improve the system result.

Social Network Approach. People place their trust on people who are closer to them. Other reviewers' feedback will be accepted based on the relationship among the user and the reviewers. Social Network Analysis (SNA) was applied in this area to derived new features to improve the RS. However, there are different terms for friendship strength being used by researchers. They are tie strength, intimacy and trust [32]. Xue *et al.* [43] had proposed a group recommendation that integrated financial social networks and collaborative filtering algorithm. Seo *et al.* [32] proposed a friendship strength based personalized RS that utilized the closeness between users on Twitter to calculate users' weight to apply on the RS. Jiang *et al.* [14] proposed a trust-based collaborative filtering algorithm that improved the performance of traditional Slope One algorithm [18] by taking the trust relationship among users and the degree of trust for ratings into consideration. Wang [40] proposed a trust-based prediction approach for RS that calculated the quantifying trust values among users to alleviate the cold start and data sparsity problems. Yuji [47] proposed a trust prediction method for RS that implemented improved Dempster-Shafer [6] Theory (DST).

4.7 Deep Learning Approach

RS has been improved by the application of deep learning algorithms. It has been proved in lots of researches. Deep Learning algorithms have four main characteristics that fitted the RS requirements for performance improvement [50].

Nonlinear Transformation. Deep learning technique is able to model the nonlinearity in data with non-linear activation using functions such as the Rectified Linear Unit (ReLU) or Sigmoid. This property makes it possible to capture the complicated user-item interaction patterns. Conventional methods are able to deal with complicated interaction patterns and accurately reflect the user's interests. Zhang *et al.* [51] proposed utilizing convolution deep learning model based on label weight nearest neighbor to obtain the nearest neighbour set that has biggest impact to the target user. Purkaystha *et al.* [31] proposed a Nonlinear User and Item Factorization (NUIF) model that it learn the complex nonlinear relationship among the user and item factors. Yang*et al.* [44] proposed a RS using Restricted Boltzmann Machines with implicit feedback that able to predict preference for a new user and achieved significant improvements.

Representation Learning. Deep learning reduces the efforts on feature engineering and allows the recommendations system to include heterogeneous features extracted from text, images, video and audio. Jiang *et al.* [15] proposed an improved algorithm based on deep neural network on measure the similarity between songs. Zhang *et al.* [48] proposed a framework called convolutional collaborative filtering with attention (Att-ConvCF) that applied attention mechanism on traditional collaborative filtering and assigned proper weights to the prediction model in order to improve feature extraction. Low *et al.* [22] proposed an convolutional neural network-based (CNN) collaborative filtering RS that model the user-item interactions and integrated it with matrix factorization. Wei *et al.* [41] proposed a deep learning based RS for cold start items that utilized stacked de-noising autoencoder to obtain item features and then integrate it with time-aware collaborative filtering model (timeSVD++). Chang *et al.* [4] proposed a personalized music recommendation system that utilized CNNs approach to classify music based on the audio signal beat into different genres and then integrated it with collaborative filtering to generate better results.

Sequence Modeling. Sequence modeling is able to mine the temporal dynamics of user preferences and item transformation. Zhang *et al.* [49] proposed a next item recommendation system by using self-attention that included short-term and long-term users' intention in model building.

Flexibility. Deep learning allows combining different structures of neural networks to form a powerful hybrid models or module replacement. Mudda *et al.* [25] proposed a S-DEEPREC Model that utilized two neural network to recommend new locations with geographical constraint consideration. Taheri *et al.* [35] proposed a deep learning based framework that unified two deep neural networks for retrieving and ranking items.

4.8   Hybrid Approach

In order to improve RS, researchers combine multiple approaches that were mentioned. Alahmadi *et al.* [1] have proposed a recommendation system that it calculate the trustworthiness between users by using the users' action on social media, Twitter. Geng *et al.* [8] proposed a two-step personalized location recommendation based on multi-objective immune algorithm (MOIA). Users' location based social network information were integrated by using multi-objective immune algorithm in the proposed framework. Parvin *et al.* [28] proposed a trust-aware collaborative filtering method based on ant colony optimization (TCF-ACO). Fressato *et al.* [29] proposed Item Most Similar based on Matrix Factorization (Item-MSMF) technique. Guo *et al.* [10] proposed a multi-view clustering method by utilizing both the view of users' social trust relationships and the view of the users' ratings. Bobadilla *et al.* [3] proposed a RS clustering using Bayesian non Negative Matrix Factorization that provided clustering and performance improvements.

5   Approaches Summary

In our review on publications over recent years (Fig. 1), most of the work on RS are focused on performance improvement (Fig. 2). The research works related to the challenges listed in Sect. 3 were discussed only in few publications. Scalability is rarely discussed among the published research works. However, there are some work addressing solely on data sparsity (Fig. 3) recently.

After the Netflix Prize competition, matrix factorization became a popular direction for RS. Many researchers worked on matrix factorization improvement. Researchers applied sentiment analysis on comments and analyzed the users social networks to gain extra features to boost the accuracy of the RS. With the recent popularity of deep learning algorithms, researchers also embarked on implementing deep learning to improve RS performance. Work on deep learning and matrix factorization accounts for more than 40% of the recent developments in RS (Fig. 4).

6   Future Research Direction

Existing works have formed a strong foundation for RS, but there are some open issues. From the brief description of the directions (Sect. 5), we focus on potential future research directions that are worth considering.

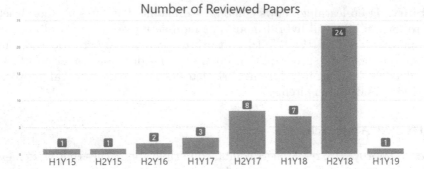

Fig. 1. Reviewed publications by published year

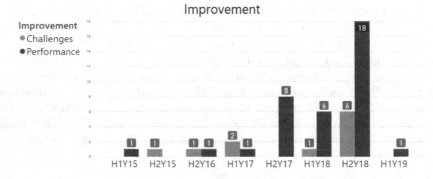

Fig. 2. Reviewed publications by published year and improvement area

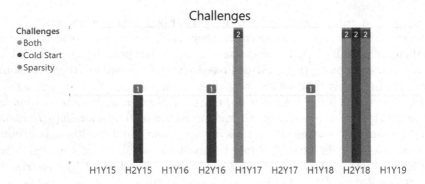

Fig. 3. Reviewed publications by published year and challenges breakdown

6.1   Recommendation System Approaches

With deep learning and matrix factorization being the strongest trends recently (Fig. 5), we expect this trend to lead the revival of RS research with many different opportunities abound.

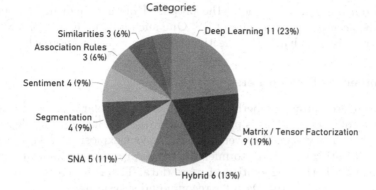

Fig. 4. Recommender Systems approaches as per reviewed articles

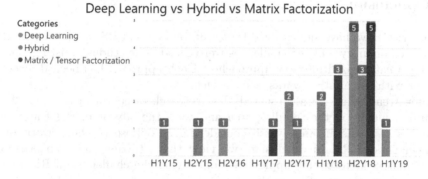

Fig. 5. Reviewed publications by published year and RS technique categories

Deep Learning Perspective. Deep learning methods are becoming more mature. Existing works have utilized deep learning on mining user profiles, similar items, review text, contextual information and implicit feedback. However, there are more information that is available and can be utilized. A more comprehensive utilization of the information may be advantageous for RS accuracy.

Feature engineering using deep learning is an area that has not been fully explored for RS. Deep learning can improve the current feature engineering process which are manually feature crafting or existing features selection. Deep neural network can be used to automate the feature crafting. Representation learning of deep neural network is able to turn the raw items (such as image, audio, video and text) to become useful features with simple feature engineering pipelines. The application of deep learning brings more opportunities in recommending various items with unstructured data like text, image, audio and video.

Matrix Factorization Perspective. There have been several publications that have improved matrix factorization. Most of the methods are independent and from our observation, some of the proposed methods can be combined to improve the accuracy performance of RS. One such area is the pre-processing of the input matrix to enhance the effectiveness of an existing approach.

6.2 Challenges Solving Perspective

Challenges related proposed method has not been a popular topic as it is a more difficult area of research. This is predominantly due to a lack of suitable relevant data and the privacy acts. However, with the recent explosion of data, it would be possible to utilize Big Data to infer the necessary data without any privacy infringement for better aggregating of users data. This is potentially an area of innovative utilization of Big Data for recommender systems.

7   Conclusion

In this article, we have summarized the common types of RS in the current RS research community and the challenges were faced. The studied existing works have been categorized into few approaches. Each approach has been discussed together with the existing works as examples.

Some trends have been presented. The research direction in the next 2 to 3 years should be in the trending area and with the advent of Big Data, the challenges of RS is an area worth considering as oppose to the performance enhancements of RS. In conclusion, we view that utilizing the deep learning and/or matrix factorization approaches for solving the challenges of RS would be an appropriate area of research.

References

1. Alahmadi, D.H., Zeng, X.: Twitter-based recommender system to address cold-start: a genetic algorithm based trust modelling and probabilistic sentiment analysis. In: 2015 IEEE 27th International Conference on Tools with Artificial Intelligence (ICTAI), pp. 1045–1052, November 2015
2. Banerjee, H., et al.: Movie recommendation system using particle swarm optimization. In: 2017 8th Annual Industrial Automation and Electromechanical Engineering Conference (IEMECON), pp. 121–126, August 2017
3. Bobadilla, J., Bojorque, R., Hernando Esteban, A., Hurtado, R.: Recommender systems clustering using bayesian non negative matrix factorization. IEEE Access **6**, 3549–3564 (2018)
4. Chang, S., Abdul, A., Chen, J., Liao, H.: A personalized music recommendation system using convolutional neural networks approach. In: 2018 IEEE International Conference on Applied System Invention (ICASI), pp. 47–49, April 2018
5. Convertini, N., Logrillo, N., Manca, F., Palmisano, T.: Recommendation system using hybrid fuzzy association rules for human smart cities. In: 2018 AEIT International Annual Conference, pp. 1–5, October 2018
6. Dempster, A.P.: Upper and lower probabilities induced by a multivalued mapping. Ann. Math. Stat. **38**, 325–339 (1967)

7. Feng, X., Yu, W., Li, Y.: Faster matrix completion using randomized SVD. In: 2018 IEEE 30th International Conference on Tools with Artificial Intelligence (ICTAI), pp. 608–615, November 2018

8. Geng, B., Jiao, L., Gong, M., Li, L., Wu, Y.: A two-step personalized location recommendation based on multi-objective immune algorithm. Inf. Sci. **475**, 161–181 (2019)

9. Gouvert, O., Oberlin, T., Févotte, C.: Negative binomial matrix factorization for recommender systems. CoRR abs/1801.01708 (2018)

10. Guo, G., Zhang, J., Yorke-Smith, N.: Leveraging multiviews of trust and similarity to enhance clustering-based recommender systems. Knowl. Based Syst. **74**, 14–27 (2015)

11. Han, C., Lin, B.: A hybrid model of tensor factorization and sentiment utility logistic model for trip recommendation. In: 2018 1st IEEE International Conference on Knowledge Innovation and Invention (ICKII), pp. 158–161, July 2018

12. Hosseinzadeh Aghdam, M.: Context-aware recommender systems using hierarchical hidden Markov model. Physica A **518**, 89–98 (2019)

13. Jain, S., Grover, A., Thakur, P.S., Choudhary, S.K.: Trends, problems and solutions of recommender system. In: International Conference on Computing, Communication Automation, pp. 955–958, May 2015

14. Jiang, L., Cheng, Y., Yang, L., Li, J., Yan, H., Wang, X.: A trust-based collaborative filtering algorithm for e-commerce recommendation system. J. Ambient Intell. Hum. Comput. (2018)

15. Jiang, M., Yang, Z., Zhao, C.: What to play next? A RNN-based music recommendation system. In: 2017 51st Asilomar Conference on Signals, Systems, and Computers, pp. 356–358, October 2017

16. Koren, Y.: Factorization meets the neighborhood: a multifaceted collaborative filtering model. In: Proceedings of the 14th ACM SIGKDD International Conference on Knowledge Discovery and Data Mining, KDD 2008, pp. 426–434. ACM, New York (2008)

17. Lee, J., Hwang, W., Parc, J., Lee, Y., Kim, S., Lee, D.: l injection: toward effective collaborative filtering using uninteresting items. IEEE Trans. Knowl. Data Eng. **31**(1), 3–16 (2019)

18. Lemire, D., Maclachlan, A.: Slope one predictors for online rating-based collaborative filtering. In: Proceedings of the 2005 SIAM International Conference on Data Mining, pp. 471–475. SIAM (2005)

19. Li, G., Zhang, J.: Music personalized recommendation system based on improved KNN algorithm. In: 2018 IEEE 3rd Advanced Information Technology, Electronic and Automation Control Conference (IAEAC), pp. 777–781, October 2018

20. Logesh, R., Subramaniyaswamy, V., Malathi, D., Sivaramakrishnan, N., Vijayakumar, V.: Enhancing recommendation stability of collaborative filtering recommender system through bio-inspired clustering ensemble method. Neural Comput. Appl. (2018)

21. Logesh, R., Subramaniyaswamy, V., Vijayakumar, V., Gao, X.Z., Indragandhi, V.: A hybrid quantum-induced swarm intelligence clustering for the urban trip recommendation in smart city. Future Gener. Comput. Syst. **83**, 653–673 (2018)

22. Low, Y.H., Yap, W.-S., Tee, Y.K.: Convolutional neural network-based collaborative filtering for recommendation systems. In: Kim, J.-H., Myung, H., Lee, S.-M. (eds.) RiTA 2018. CCIS, vol. 1015, pp. 117–131. Springer, Singapore (2019). https://doi.org/10.1007/978-981-13-7780-8_10

23. Luo, L., Xie, H., Rao, Y., Wang, F.L.: Personalized recommendation by matrix co-factorization with tags and time information. Expert Syst. Appl. **119**, 311–321 (2019)
24. Mu, Y., Xiao, N., Tang, R., Luo, L., Yin, X.: An efficient similarity measure for collaborative filtering. Procedia Comput. Sci. **147**, 416–421 (2019). 2018 International Conference on Identification, Information and Knowledge in the Internet of Things
25. Mudda, S., Lian, D., Giordano, S., Liu, D., Xie, X.: Spatial-aware deep recommender system. In: 2018 IEEE SmartWorld, Ubiquitous Intelligence Computing, Advanced Trusted Computing, Scalable Computing Communications, Cloud Big Data Computing, Internet of People and Smart City Innovation (SmartWorld/SCALCOM/UIC/ATC/CBDCom/IOP/SCI), pp. 983–990, October 2018
26. Nabil, S., Elbouhdidi, J., Yassin, M.: Recommendation system based on data analysis-application on tweets sentiment analysis. In: 2018 IEEE 5th International Congress on Information Science and Technology (CiSt), pp. 155–160, October 2018
27. Osadchiy, T., Poliakov, I., Olivier, P., Rowland, M., Foster, E.: Recommender system based on pairwise association rules. Expert Syst. Appl. **115**, 535–542 (2019)
28. Parvin, H., Moradi, P., Esmaeili, S.: TCFACO: trust-aware collaborative filtering method based on ant colony optimization. Expert Syst. Appl. **118**, 152–168 (2019)
29. Pereira Fressato, E., Fortes da Costa, A., Garcia Manzato, M.: Similarity-based matrix factorization for item cold-start in recommender systems. In: 2018 7th Brazilian Conference on Intelligent Systems (BRACIS), pp. 342–347, October 2018
30. Preethi, G., Krishna, P.V., Obaidat, M.S., Saritha, V., Yenduri, S.: Application of deep learning to sentiment analysis for recommender system on cloud. In: 2017 International Conference on Computer, Information and Telecommunication Systems (CITS), pp. 93–97, July 2017
31. Purkaystha, B., Datta, T., Islam, M.S.: Marium-E-Jannat: rating prediction for recommendation: constructing user profiles and item characteristics using back-propagation. Appl. Soft Comput. **75**, 310–322 (2019)
32. Seo, Y.D., Kim, Y.G., Lee, E., Baik, D.K.: Personalized recommender system based on friendship strength in social network services. Expert Syst. Appl. **69**, 135–148 (2017)
33. Subramaniyaswamy, V., Logesh, R.: Adaptive KNN based recommender system through mining of user preferences. Wireless Pers. Commun. **97**(2), 2229–2247 (2017)
34. Symeonidis, P., Malakoudis, D.: Multi-modal matrix factorization with side information for recommending massive open online courses. Expert Syst. Appl. **118**, 261–271 (2019)
35. Taheri, S.M., Irajian, I.: DeepMovRS: a unified framework for deep learning-based movie recommender systems. In: 2018 6th Iranian Joint Congress on Fuzzy and Intelligent Systems (CFIS), pp. 200–204, February 2018
36. Tahmasbi, H., Jalali, M., Shakeri, H.: Modeling temporal dynamics of user preferences in movie recommendation. In: 2018 8th International Conference on Computer and Knowledge Engineering (ICCKE), pp. 194–199, October 2018
37. Wang, J., Liu, T.: Improving sentiment rating of movie review comments for recommendation. In: 2017 IEEE International Conference on Consumer Electronics - Taiwan (ICCE-TW), pp. 433–434, June 2017
38. Wang, K., Jin, Y., Wang, H., Peng, H., Wang, X.: Personalized time-aware tag recommendation. In: AAAI (2018)

39. Wang, M., Xiao, Y., Zheng, W., Jiao, X.: RNDM: a random walk method for music recommendation by considering novelty, diversity, and mainstream. In: 2018 IEEE 30th International Conference on Tools with Artificial Intelligence (ICTAI), pp. 177–183, November 2018
40. Wang, P., Huang, H., Zhu, J., Qi, L.: A trust-based prediction approach for recommendation system. In: Yang, A., et al. (eds.) SERVICES 2018. LNCS, vol. 10975, pp. 157–164. Springer, Cham (2018). https://doi.org/10.1007/978-3-319-94472-2_12
41. Wei, J., He, J., Chen, K., Zhou, Y., Tang, Z.: Collaborative filtering and deep learning based recommendation system for cold start items. Expert Syst. Appl. **69**, 29–39 (2017)
42. Xu, A.L., Liu, B.J., Gu, C.Y.: A recommendation system based on extreme gradient boosting classifier. In: 2018 10th International Conference on Modelling, Identification and Control (ICMIC), pp. 1–5, July 2018
43. Xue, J., Zhu, E., Liu, Q., Yin, J.: Group recommendation based on financial social network for robo-advisor. IEEE Access **6**, 54527–54535 (2018)
44. Yang, F., Lu, Y.: Restricted Boltzmann machines for recommender systems with implicit feedback. In: 2018 IEEE International Conference on Big Data (Big Data), pp. 4109–4113, December 2018
45. Yang, W., Fan, S., Wang, H.: An item-diversity-based collaborative filtering algorithm to improve the accuracy of recommender system. In: 2018 IEEE SmartWorld, Ubiquitous Intelligence Computing, Advanced Trusted Computing, Scalable Computing Communications, Cloud Big Data Computing, Internet of People and Smart City Innovation (SmartWorld/SCALCOM/UIC/ATC/CBDCom/IOP/SCI), pp. 106–110, October 2018
46. Yi, K., Chen, T., Cong, G.: Library personalized recommendation service method based on improved association rules. Library Hi Tech **36**(3), 443–457 (2018)
47. Yuji, W.: A trust prediction method for recommendation system. In: 2017 9th International Conference on Intelligent Human-Machine Systems and Cybernetics (IHMSC), vol. 2, pp. 64 68, August 2017
48. Zhang, B., Zhang, H., Sun, X., Feng, G., He, C.: Integrating an attention mechanism and convolution collaborative filtering for document context-aware rating prediction. IEEE Access **7**, 3826 3835 (2019)
49. Zhang, S., Tay, Y., Yao, L., Sun, A.: Next item recommendation with self-attention. CoRR abs/1808.06414 (2018)
50. Zhang, S., Yao, L., Sun, A.: Deep learning based recommender system: a survey and new perspectives. CoRR abs/1707.07435 (2017)
51. Zhang, W., Liu, F., Jiang, L., Xu, D.: Recommendation based on collaborative filtering by convolution deep learning model based on label weight nearest neighbor. In: 2017 10th International Symposium on Computational Intelligence and Design (ISCID), vol. 2, pp. 504–507, December 2017
52. Zheng, D., Xiong, Y.: A unified probabilistic matrix factorization recommendation algorithm. In: 2018 International Conference on Robots Intelligent System (ICRIS), pp. 246–249, May 2018

Effect of Feature Selection in Software Fault Detection

Shamse Tasnim Cynthia, Md. Golam Rasul, and Shamim Ripon[⊠]

Department of Computer Science and Engineering, East West University,
Dhaka, Bangladesh
cnth999@gmail.com, grpranto@gmail.com, dshr@ewubd.edu

Abstract. The quality of software is enormously affected by the faults associated with it. Detection of faults at a proper stage in software development is a challenging task and plays a vital role in the quality of the software. Machine learning is, now a days, a commonly used technique for fault detection and prediction. However, the effectiveness of the fault detection mechanism is impacted by the number of attributes in the publicly available datasets. Feature selection is the process of selecting a subset of all the features that are most influential to the classification and it is a challenging task. This paper thoroughly investigates the effect of various feature selection techniques on software fault classification by using NASA's some benchmark publicly available datasets. Various metrics are used to analyze the performance of the feature selection techniques. The experiment discovers that the most important and relevant features can be selected by the adopted feature selection techniques without sacrificing the performance of fault detection.

Keywords: Fault detection · Feature selection · Feature classification

1 Introduction

Nowadays the role of software has gained much more importance in every known field. This makes the software system more complex than before and some of the software systems have to be delivered with the least or non-negligible number of faults possible. Faults are basically the errors in the code that prevents the software system to work as expected [6]. Faults are often generated for misunderstanding, lacking knowledge in the working area or even for the deadlines. Some of the software faults can be detected at the early stage of developing a software but when fault is detected in the later stage, not only it takes a lot of time to fix the faults but also the quality of software is compromised. Several studies reveal that almost 80% of the software faults occur from 20% of the modules and defect free portion covers the rest half of the module [29]. Software faults demand the rework process that has negative impact on for Software Quality Assurance [10]. Ability in detecting software faults mostly assist software developers in the testing phase about maintaining software standards [8]. So predicting software faults

© Springer Nature Switzerland AG 2019
R. Chamchong and K. W. Wong (Eds.): MIWAI 2019, LNAI 11909, pp. 52–63, 2019.
https://doi.org/10.1007/978-3-030-33709-4_5

at the early stage of software development can support the development of more efficient and reliable software within the stipulated limited time and cost [18].

Feature selection is one of the most significant techniques for kind of any data analysis. Features are mainly referred as the attributes which are given in a dataset and have strong correlation with the class attributes [4]. The main aim of feature selection in a dataset is to select the essential features which will help to improve the fruitfulness of a model. The feature selection techniques not only increases the accuracy and efficiency of a classifier but also decreases the chance of overfitting, reduces the dimensionality and eliminates noise [11,21]. In addition, it can seek out useful information deliberately and lessens the effect of variance in the result. Proper selection of features can help researchers looking for the exact fault in the model. Again when the most essential features are selected, the reduced dimensionality of a dataset boosts the performance of some algorithms, delivers more accurate results in the less amount of time. Most of the feature selection techniques extract features that ranges from sub-optimal to near optimal solutions [26]. By ranking the features with different scores, feature selection techniques reach to near optimal solutions.

Considering the significance of feature selection in predicting and classifying software faults, this work aims at investigating the effect of various feature selection techniques upon the performance of various classification algorithms. In particular, several feature selection techniques are applied to some used fault prediction datasets and then apply classification and predictive techniques on the datasets having only those features selected earlier.

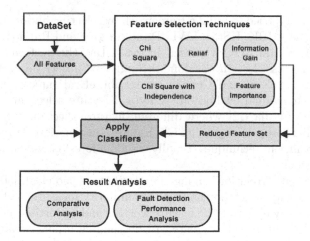

Fig. 1. Proposed model

The schematic view of the proposed framework is illustrated in Fig. 1. Five feature selection techniques along with five datasets are considered in the framework for experimental purpose. For each dataset, all the feature selection techniques are applied and relevant features are selected for each type of technique.

Classification algorithms are then applied to the selected features obtained from each feature selection technique. Several classification algorithms are applied in the experiments. Various metrics are used to analyze the performance of each classification algorithm for each type of feature selection technique and for each type of dataset. For comparative analysis, experiment has also been conducted considering all the features in the dataset and then compare the result with that of the selected features.

The rest of the paper is organized as follows. A brief overview of the dataset used in the paper is shown in Sect. 3. After briefly demonstrated the adopted feature selection techniques in Sect. 4, the following section shows the effect of feature selection on the applied classification algorithms for various datasets. A thorough analysis of the obtained result is presented in Sect. 6. A brief review of similar works are described in Sect. 2. Finally, Sect. 7 concludes the paper by summarizing the paper and outlining our future plan.

2   Related Work

A framework model has been proposed by Oinbao Song et al. [25] to follow-up the MGF on defect prediction using Scheme evaluation and defect prediction for feature selection. Naive Base, J4.8 and OneR were used for comparing the performance. Jiang et al. [13] used ROCUS for software defect prediction. They proposed a disagreement-based semi-supervised learning method to exploit the abundant unlabeled data but higher misclassification rate is the limitation for this technique.

Kakkar et al. [14] tried to build a framework by selecting important attributes using five classifiers: IBK, KStar, LWL, Random Tree and Random Forest. The values of accuracy and ROC was used in evaluating the performance of these classifiers. Attributes selection was done through CfxSubsetEval evaluator where ChiSquaredAttributeEval and CorrelationAttributeEval ranked the attributes based on their individual evaluation. A hybrid feature selection approach has been introduced by Jia [12] where different feature selection techniques have been combined. Those techniques are Chi Square, Information Gain and Pearson Correlation Coefficient techniques. Finally, random forest classifier has been used to build the model.

To measure the correlation among the attributes Qiao et al. [30] introduced a feature selection approach on the basis of similarity measurement for software defect prediction. By updating the feature weights and by sorting them according to their rankings feature list is created in descending order. Finally, K-nearest model classifier is applied on the selected dataset for the detection of faults. Xu et al. [28] proposed a feature selection framework named MICHAC which stands for Maximal Information Coefficient with Hierarchical Agglomerative Clustering. To remove irrelevant features, this framework extracts one feature from each feature subset groups. Three different classifiers and four performance metrics were used to evaluate the performance of the model built with selected featured datasets.

Ibrahim et al. [10] in their work, used Bat-Based Search Algorithm for the feature selection purpose. Similarly, Wahono et al. in their research [26] on software defect prediction gave priority on imbalance nature of the NASA dataset and for feature selection, Genetic Algorithm has been used only. Anbu et al. in their research [2] used Firefly algorithm for feature selection and classifiers like Support Vector Machine, Naïve Bayes and K- nearest neighbor to predict the defects.

In comparison to the existing works, in this paper five feature selection techniques have been applied to NASA MDP's five datasets and five search-based classifiers are applied for analysis of classification performance. Combination of all these three factors results in a large number experiments. Such experiment can give better intuition regarding the effectiveness of the feature selection techniques.

3   DataSet Overview

NASA MDP dataset [26] has been used in this paper. This dataset contains 96 datasets, but among them only 13 datasets are provided by NASA [22]. For experimental purpose we have taken only five of them. These datasets are collected from several projects (satellite instrumentation, ground control systems, attitude control system etc.) in USA for several years. Each dataset is the representative of NASA software system/subsystem containing metrics of static code and fault data for each comprising module. These datasets have been used very widely for detecting software fault. The fault prediction dataset has McCabe, Halstead and line of code metrics [3]. The attributes of these datasets are mostly of numeric types except the class attribute which consists polynomial data. Table 1 shows the total instances of the selected datasets and their class percentages.

Table 1. Dataset Overview

Dataset	Total sample	Defective	Not defective	Number of attributes	Programming language
CM1	344	42 (12.2%)	302 (88%)	37	C
KC3	200	36 (18%)	164 (82%)	39	Java
PC4	1399	178 (13%)	1221 (87%)	37	C
PC2	1585	16 (1%)	1569 (99%)	37	C
MW1	264	27 (10%)	237 (90%)	37	C

4   Feature Selection Techniques

The *Chi-Square* test is introduced by Karl Pearson is a statistical hypothesis test that determines the goodness of fit between a set of observed and expected values [5]. It is a nonparametric test that is used for testing the hypothesis of no association between two or more groups, population or criteria and to test how well the observed distribution of data fits with the distribution that is expected [23]. The formula of Chi-square is shown in Eq. 1

$$\chi_c^2 = \sum_1^n \frac{(O_i - E_i)^2}{E_i} \tag{1}$$

where, c is degree of freedom, O is observed value and E is expected value

The *Chi-square test of independence* is used to detect if there is a significant relationship between two nominal (categorical) variables. Each category's frequency for one nominal variable is compared with the categories of another nominal variable. Each row of the data in a contingency table represents a category for one variable and each column represents a category for other variable [16]. The corresponding formula is show in Eq. 2,

$$\chi^2 = \sum_{i=1}^r \sum_{j=1}^c \frac{(O_{ij} - E_{ij})^2}{E_{ij}} \tag{2}$$

where, r is number of rows, c is number of rows, O is observed value and E is expected value.

Information Gain is a measure of the change of entropy which reduces the uncertainty of the result. Entropy gives the measure of impurity of the classes. The value of the entropy should be less for getting the best output. When a node in a decision tree is used for partitioning the training instances into smaller subsets, the value of the entropy changes. Information gain specifies the importance of an attribute and decides the ordering of the attributes in the nodes of a Decision Tree.

$$Gain(T, x) = Entropy(T) - Entropy(T, x) \tag{3}$$

Relief is a feature selection algorithm which uses a statistical method and avoids heuristic research. The algorithm inspired by instance-based learning. It needs linear time for the number of given features and the number of training instances regardless of the target concept to be learned.

From given training data, sample size, and a threshold of relevancy, Relief finds those features that are statistically relevant to the target concept. Relief collects the total number of triplets of an instance, its Near-hit instance and Near-miss instance. Euclidian distance is used for selecting Near-hit and Near-miss. A routine is also called by Relief to update feature weight vector for every triplet and finds the average feature weight vector Relevance (of all the features to the target concept) and those features whose average weight is above the given threshold are selected by Relief [15].

Feature Importance returns a score for each feature and based on that score, the features which have higher score get more privilege towards the output variable. It uses ensembles of decision trees which computes the relative importance of each attribute.

5   Effect of Feature Selection

We have evaluated the performance of our five feature selection processes using the True Positive Rate, True Negative Rate, Precision and Accuracy. These metrics help us to examine whether the methods can correctly and efficiently recognize the optimized features and show us the effects of feature selection in the classification [1]. The experimental result shown here are obtained by considering 20 features selected by applying the feature selection techniques mentioned here.

In the experiment of feature selection techniques with classification algorithms, various matrices are used to measure the performance, namely TPR and TNR and *accuracy*. Four important information is obtained from confusion matrix to calculate these matrices: True Positive (TP), True Negative (TN), False Positive (FP) and False Negative (FN). True positive rate or Sensitivity is the result where the positive class is correctly predicted by the model.

$$TPR = \frac{TP}{TP + FN} \tag{4}$$

Similarly, true negative rate or Specificity is the result where the negative class is correctly predicted by the model.

$$TNR = \frac{TN}{TN + FP} \tag{5}$$

Classification accuracy is the fraction of prediction to see whether the model works right.

$$Accuracy = \frac{TP + TN}{TP + TN + FP + FN} \tag{6}$$

True positive rate, true negative rate and accuracy these three metrics need to be higher for better prediction. We have calculated all these three metrics on the five datasets using some selected classifiers to see their performances. All these experiments have been conducted considering not only the selected feature but also for all the features. While choosing the classifiers, only search-based classifiers as mentioned in [24] are selected for experimental purpose. The classifiers used here are: Decision Tree [20], Random Forest [7], Naïve Bayes [27], Logistic Regression [19] and Artificial Neural Network [17].

Table 2 illustrates True Positive Rate (TPR), True Negative Rate (TNR) and Accuracy of the the classifiers over the datasets after choosing the features by using Relief test. Table 3, on the other hand illustrates the TPR, TNR and Accuracy values of different datasets where features are selected through Chi Square test. Tables 4, 5, and 6 shows the same for Information gain, Chi Square Test of Independence and Feature Importance respectively. For this experiment, 20 relevant features have been selected from each dataset by using the feature selection techniques. We have also conducted a similar experiment by considering 10 most relevant features. The F-measure and precision are also calculated for all the datasets considering all the algorithms and feature selection techniques. The comparative analysis of performance between 10 features and 20 features is illustrated in Sect. 6.

Table 2. Performance with relief test

Dataset	Decision Tree			Random Forest			Naïve Bayes			Logistic Regression			ANN		
	TPR	TNR	Accuracy	TPR	TNR	Accuracy	TPR	TNR	Accuracy	TPR	TNR	Accuracy	TPR	TNR	Accuracy
CM1	0.00	1.00	88%	0.38	1.00	92%	0.29	0.90	83%	0.26	0.98	89%	0.10	0.99	88%
KC3	0.56	1.00	83%	0.61	1.00	93%	0.39	0.89	81%	0.36	0.96	86%	0.14	0.98	83%
PC2	0.38	1.00	99%	0.88	1.00	99%	0.31	0.96	95%	0.13	1.00	99%	0.00	1.00	99%
PC4	0.35	0.99	91%	0.23	1.00	91%	0.56	0.86	82%	0.38	0.98	90%	0.48	0.98	92%
MW1	0.00	1.00	90%	0.63	1.00	96%	0.56	0.85	82%	0.40	0.98	91%	0.30	0.98	91%

Table 3. Performance with Chi-Square

Dataset	Decision Tree			Random Forest			Naïve Bayes			Logistic Regression			ANN		
	TPR	TNR	Accuracy	TPR	TNR	Accuracy	TPR	TNR	Accuracy	TPR	TNR	Accuracy	TPR	TNR	Accuracy
CM1	0.43	0.99	92%	0.33	1.00	92%	0.29	0.90	83%	0.24	0.97	87%	0.09	0.99	88%
KC3	0.06	1.00	83%	0.67	1.00	94%	0.36	0.91	81%	0.31	0.96	84%	0.11	0.98	83%
PC2	0.25	1.00	99%	0.88	1.00	99%	0.19	0.97	96%	0.19	0.99	99%	0.00	1.00	99%
PC4	0.29	0.99	91%	0.25	1.00	90%	0.26	0.94	85%	0.39	0.98	91%	0.41	0.98	91%
MW1	0.11	1.00	91%	0.67	1.00	97%	0.56	0.85	82%	0.33	0.99	92%	0.44	0.99	93%

Table 4. Performance of information gain

Dataset	Decision Tree			Random Forest			Naïve Bayes			Logistic Regression			ANN		
	TPR	TNR	Accuracy	TPR	TNR	Accuracy	TPR	TNR	Accuracy	TPR	TNR	Accuracy	TPR	TNR	Accuracy
CM1	0.38	0.99	92%	0.36	1.00	92%	0.36	0.90	84%	0.26	0.99	90%	0.17	0.99	90%
KC3	0.50	0.99	90%	0.64	0.99	93%	0.34	0.91	82%	0.44	0.97	88%	0.31	1.00	87%
PC2	0.25	1.00	99%	0.69	1.00	99%	0.19	0.97	96%	0.19	1.00	99%	0.00	1.00	99%
PC4	0.30	0.99	90%	0.26	1.00	90%	0.54	0.92	88%	0.43	0.99	91%	0.51	0.98	92%
MW1	0.11	1.00	91%	0.59	1.00	96%	0.59	0.86	83%	0.33	0.99	93%	0.44	0.99	94%

Table 5. Performance of Chi Square test of independence

Dataset	Decision Tree			Random Forest			Naïve Bayes			Logistic Regression			ANN		
	TPR	TNR	Accuracy	TPR	TNR	Accuracy	TPR	TNR	Accuracy	TPR	TNR	Accuracy	TPR	TNR	Accuracy
CM1	0.43	0.99	92%	0.33	1.00	92%	0.29	0.90	83%	0.24	0.97	89%	0.09	0.99	88%
KC3	0.06	1.00	83%	0.67	1.00	94%	0.36	0.91	81%	0.31	0.96	84%	0.11	0.98	82%
PC2	0.25	1.00	99%	0.88	1.00	99%	0.19	0.97	96%	0.19	0.99	99%	0	1	99%
PC4	0.29	0.99	91%	0.25	1.00	90%	0.26	0.94	85%	0.39	0.98	91%	0.41	0.98	91%
MW1	0.11	1.00	91%	0.63	1.00	96%	0.51	0.99	83%	0.52	0.87	93%	0.44	0.98	93%

Table 6. Performance of feature importance

Dataset	Decision Tree			Random Forest			Naïve			Logistic Regression			ANN		
	TPR	TNR	Accuracy	TPR	TNR	Accuracy	TPR	TNR	Accuracy	TPR	TNR	Accuracy	TPR	TNR	Accuracy
CM1	0.43	0.99	92%	0.38	1.00	92%	0.38	0.89	83%	0.26	0.98	90%	0.14	0.98	88%
KC3	0.06	1.00	83%	0.67	1.00	94%	0.36	0.91	82%	0.42	0.96	87%	0.22	0.98	85%
PC2	0.31	1.00	99%	0.88	1.00	99%	0.19	0.97	96%	0.25	0.99	99%	0.00	1.00	99%
PC4	0.30	0.99	91%	0.30	1.00	91%	0.30	0.94	85%	0.40	0.98	91%	0.43	0.98	91%
MW1	0.11	1.00	92%	0.67	1.00	97%	0.56	0.86	83%	0.33	0.99	92%	0.41	0.98	93%

6   Result Analysis

The aim of feature selection is to find the features which are more important and relevant to the target class and discard those which are less important or the correlation between them and the target class is not enough to be considered during classification. NSA dataset consisting 13 datasets, each of them has a large numbers of attributes. If it is possible to select only the important features, the computational time can be reduced, the utilization of resources can be improved and the classification efficiency can be increased. In this paper, five (5) feature selection processes are applied to select features from the datasets and only 20 most important features are selected for the classification process.

Our experiment reveals that when Decision Tree classifier is applied on CM1 dataset considering all the feature, DT can only predict N values (not defective class) and failed to predict any Y value (defective class). The Not defective class only has the class precision value. But when the same experiment is conducted by selecting features using Chi-Square Test, DT can then predict both N and Y value. The same experiment is conducted for all the five feature selection techniques for all the classifiers. The experimental result from DT is shown in Table 7. From the table it can be shown that among the five processes Relief could not predict the Defective class like the other classes.

Table 7. Class detection comparison after feature selection

	TP	TN
Chi-Square	300	18
Information gain	299	16
Feature importance	299	18
Chi-Square test of independence	300	18
Relief	302	00

On the other hand, in KC3 dataset, the feature selection process has less effect compared to the result that was calculated considering all the features. The TP and TN values found usually the same or a little different from the main dataset calculation. For the PC2 dataset, the total number of defective class is very less, so the algorithms could not work better in the classification. The Naïve Bayes and Logistic Regression algorithms perform better in detecting defective classes. The datasets with all features and the datasets with the selected features show almost the same result in both cases. In the PC4 dataset, the features selected by Relief process worked better for Naïve Bayes and Decision Tree algorithms and Chi Square test of independence performed better for Random Forest algorithm compared to the result computed when all the features were present. For example, the TP and TN values calculated with all the features are 69 and 1158 respectively, but with 20 selected features via Relief, the TP and TN values become 100 and 1049 respectively. Lastly, for the MW1 dataset, all the algorithms with all the features and with the selected features performed almost the same.

In Fig. 2 the accuracies of the different dataset are shown for the Decision tree classifier. Each dataset is tested with all the features and also with the selected features. Average values of the accuracies are taken here for comparison due to multidimensional result. Similar results are obtained for other classifiers, however due to limited scope they are omitted from here.

It is mentioned earlier that experiments have been conducted also by selecting ten (10) most relevant features in the same way as of 20 features. Figure 3 illustrates the accuracy comparison between 10 features' average accuracy and 20 features' average accuracy when random forest classifier is applied for the prediction. Here we can see that for dataset MW1, PC4, and PC2, the accuracies in both subsets of the datasets are almost same. Whereas, for KC3 the classifier gives better result 20 features and for CM1 the classifier gives better result for 10 features. For other classifiers, it has been observed that some classifiers worked better for 10 features' and some are better for 20 features' subsets but the differences are negligible.

The experiments conducted so far do not consider cross validation of the datasets. We conduct a similar experiment considering 10-fold cross validation on all the datasets by applying all five feature selection techniques using all the already chosen classifiers. As the result is multi-dimensional it cannot be presented in a single table or diagram. Figure 4 illustrates the accuracies of various classifiers on the five datasets after applying Chi-Square feature selection technique. The result shows that feature selection improves the classification accuracies for most of the datasets. This experiment has also been conducted for all the feature selection techniques to observe their performances. The resultant tables can be found in Appendix for reviewing purpose only.

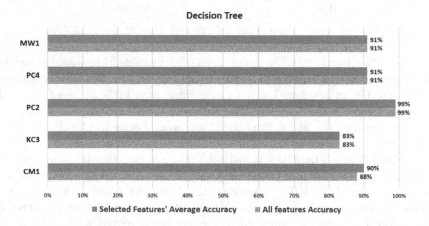

Fig. 2. Accuracy of different datasets before and after applying feature selection

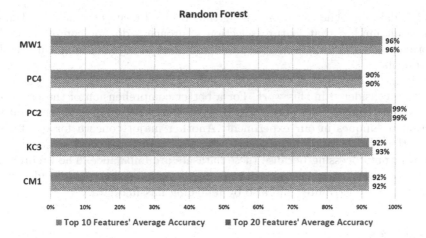

Fig. 3. Classification accuracy of random forest after applying 10 and 20 features in the datasets

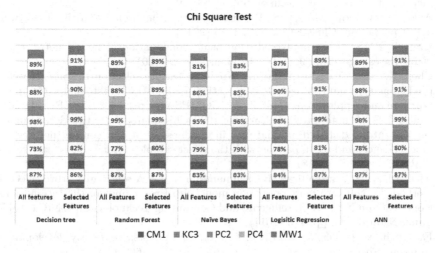

Fig. 4. Comparison of classification accuracies for all classifiers using Chi-square feature selection and 10-fold cross validation

7   Conclusion

Proper selection of relevant features in a large dataset can immensely improve the performance of classifiers and significantly reduces the training time. Among the various feature techniques, this paper shows the effect of feature selection of only five approaches. Five search-based classifiers are applied here for our experiments. The experimental results reveal that after feature selection the performance of the classifiers are almost similar to that of without feature selection. Experiments have been conducted by considering both 10 and 20 features from the datasets. The variation among the obtained results are not significant. Such result implies that feature selection approaches do not compromise the performance of the classifiers while taking less time and resource during the experiments.

It is mentioned earlier that only a subset of feature selection techniques have been considered in this work. For a better comprehension of the proposed approach, our future plan is to consider both filter and wrapper based feature selection techniques in our experiment. Another major concern for our future work is that NASA MDP datasets that we used in our experiment requires some important preprocessing because these datasets are imbalance. The preprocessing can be done using several methods [9] like eliminating module identifier, extra error data attributes and also by replacing missing data.

References

1. Agarwal, S., Tomar, D.: A feature selection based model for software defect prediction. Int. J. Adv. Sci. Technol. **65**, 39–58 (2014)
2. Anbu, M., Anandha Mala, G.S.: Feature selection using firefly algorithm in software defect prediction. Cluster Comput., 1–10 (2017)
3. Arasteh, B.: Software fault-prediction using combination of neural network and Naive Bayes algorithm. J. Netw. Technol. **9**(3), 94 (2018)
4. Chen, X., Shen, Y., Cui, Z., Ju, X.: Applying feature selection to software defect prediction using multi-objective optimization. In 2017 IEEE 41st Annual Computer Software and Applications Conference (COMPSAC), pp. 54–59. IEEE, July 2017
5. Crack, T.F.: A note on Karl Pearson's 1900 Chi-squared test: two derivations of the asymptotic distribution, and uses in goodness of fit and contingency tests of independence, and a comparison with the exact sample variance chi-square result. SSRN Electron. J. (2018)
6. Akalya Devi, C., Surendiran, B., Kannammal, K.E.: A study of feature selection methods for software fault prediction model. In: Proceedings of the International Conference on Network, Intelligence and Computing Technologies (ICNICT 2011), Tamil Nadu, India, pp. 1–5 (2011)
7. Fawagreh, K., Gaber, M.M., Elyan, E.: Random forests: from early developments to recent advancements. Syst. Sci. Control Eng. **2**(1), 602–609 (2014)
8. Felix, E.A., Lee, S.P.: Integrated approach to software defect prediction. IEEE Access **5**, 21524–21547 (2017)
9. Gray, D., Bowes, D., Davey, N., Sun, Y., Christianson, B.: The misuse of the NASA metrics data program data sets for automated software defect prediction. In: 15th Annual Conference on Evaluation & Assessment in Software Engineering (EASE 2011), pp. 96–103. IET (2011)
10. Ibrahim, D.R., Ghnemat, R., Hudaib, A.: Software defect prediction using feature selection and random forest algorithm. In: 2017 International Conference on New Trends in Computing Sciences (ICTCS), pp. 252–257. IEEE, October 2017
11. Jakhar, A.K., Rajnish, K.: Software fault prediction with data mining techniques by using feature selection based models. Int. J. Electr. Eng. Inf. **10**(3), 447–465 (2018)
12. Jia, L.: A hybrid feature selection method for software defect prediction. IOP Conf. Ser. Mater. Sci. Eng. **394**(3), 032035 (2018)

13. Jiang, Y., Li, M., Zhou, Z.-H.: Software defect detection with ROCUS. J. Comput. Sci. Technol. **26**(2), 328–342 (2011)
14. Kakkar, M., Jain, S.: Feature selection in software defect prediction: a comparative study. In 2016 6th International Conference - Cloud System and Big Data Engineering (Confluence), pp. 658–663. IEEE, January 2016
15. Kira, K., Rendell, L.A.: A practical approach to feature selection. In: Proceedings of the Ninth International Workshop on Machine Learning, pp. 249–256 (1992)
16. McHugh, M.L.: The Chi-square test of independence. Biochemia Medica, 143–149 (2013)
17. Mishra, M., Srivastava, M.: A view of artificial neural network. In: 2014 International Conference on Advances in Engineering & Technology Research (ICAETR - 2014), pp. 1–3. IEEE, August 2014
18. Nugroho, A., Chaudron, M.R.V., Arisholm, E.: Assessing UML design metrics for predicting fault-prone classes in a Java system. In: 2010 7th IEEE Working Conference on Mining Software Repositories (MSR 2010), pp. 21–30. IEEE, May 2010
19. Joanne Peng, C.-Y., Lee, K.L., Ingersoll, G.M.: An introduction to logistic regression analysis and reporting. J. Educ. Res. **96**(1), 3–14 (2002)
20. Quinlan, J.R.: Induction of decision trees. Mach. Learn. **1**(1), 81–106 (1986)
21. Rokach, L.: Ensemble-based classifiers. Artif. Intell. Rev. **33**(1–2), 1–39 (2010)
22. Shepperd, M., Song, Q., Sun, Z., Mair, C.: Data quality: some comments on the NASA software defect data sets. **2010**(9), 1–13 (2013)
23. Singhal, R., Rana, R.: Chi-square test and its application in hypothesis testing. J. Pract. Cardiovasc. Sci. **1**(1), 69 (2015)
24. Son, L.H., et al.: Empirical study of software defect prediction: a systematic mapping. Symmetry **11**(2) (2019)
25. Song, Q., Jia, Z., Shepperd, M., Ying, S., Liu, J.: A general software defect-proneness prediction framework. IEEE Trans. Software Eng. **37**(3), 356–370 (2011)
26. Wahono, R.S., Herman, N.S.: Genetic feature selection for software defect prediction. Adv. Sci. Lett. **20**(1), 239–244 (2014)
27. Webb, G.I., Keogh, E., Miikkulainen, R., Sebag, M.: Naïve Bayes. In: Sammut, C., Webb, G.I. (eds.) Encyclopedia of Machine Learning, pp. 713–714. Springer, Boston (2011). https://doi.org/10.1007/978-0-387-30164-8_576
28. Xu, Z., Xuan, J., Liu, J., Cui, X.: MICHAC: defect prediction via feature selection based on maximal information coefficient with hierarchical agglomerative clustering. In: 2016 IEEE 23rd International Conference on Software Analysis, Evolution, and Reengineering (SANER), pp. 370–381. IEEE, March 2016
29. Yousef, A.H.: Extracting software static defect models using data mining. Ain Shams Eng. J. **6**(1), 133–144 (2015)
30. Qiao, Y., Jiang, S., Wang, R., Wang, H.: A feature selection approach based on a similarity measure for software defect prediction. Front. Inf. Technol. Electron. Eng. **18**(11), 1744–1753 (2017)

An Accurate 1D Camera Calibration Based on Weighted Similar-Invariant Linear Algorithm

Lixia Lin[1], Lijun Wu[1(✉)] [ID], Songlin Lai[1], Zhicong Chen[1] [ID], Peijie Lin[1], and Zhenhui Wu[2]

[1] College of Physics and Information Engineering, Fuzhou University,
Fuzhou 350116, China
lijun.wu@fzu.edu.cn
[2] State Grid Fuzhou Electric Power Supply Company, Fuzhou 350116, China

Abstract. In recent years, researchers around the world have been researching and improving the technique of 1D calibration of cameras. The previous work has been primarily focused on reducing the motion constraints of one-dimensional calibration objects, however the accuracy of the existing methods still needs to be improved when random noise is introduces. In order to improve the accuracy of the one-dimensional calibration of the camera, in this paper, we propose a new calibration method by combining a weighted similar invariant linear algorithm and an improved nonlinear optimization algorithm. Specifically, we use the weighted similar invariant linear algorithm to obtain the camera parameters as the initial calibration parameters, and then optimize the parameters by using improved nonlinear algorithm. Finally, in the case of introducing random noise, the results of computer simulations and laboratory experiments show that when the noise level reaches 2 pixels, the parameter error of this method is mostly reduced to 0.2% compared with other methods, which verifies the feasibility of our proposed method.

Keywords: Camera calibration · Linear algorithm · Nonlinear optimization · 1D calibration objects

1 Introduction

Camera calibration is an essential step to extract metric information from 2D images in the fields of computer vision. According to the dimension of the calibration object, the existing camera calibration techniques are roughly classified into four categories: three-dimensional reference object based calibration (3D) [1–3], two-dimensional plane based calibration (2D) [4–6], one-dimensional line based calibration (1D) [7–10] and zero-dimensional calibration (0D) [11,12]. In particular, the techniques of 1D calibration are mainly applied to calibrate the relative geometric relationships between multiple cameras and the internal parameters

© Springer Nature Switzerland AG 2019
R. Chamchong and K. W. Wong (Eds.): MIWAI 2019, LNAI 11909, pp. 64–75, 2019.
https://doi.org/10.1007/978-3-030-33709-4_6

of each individual camera. The techniques of 0D calibration require the multiple parameter estimations and hence involve many complicated mathematical problems. Therefore, considering the Euclidean information of collinearity and distance between the markers on the calibration object, the techniques of 1D calibration are superior to the 0D calibration in the terms of algorithm complexity, stability and accuracy. Compared to the 2D and 3D calibration methods, the 1D calibration method makes it easier to construct the calibration because the geometry required for the 1D calibration object has been reduced to a 1D object with at least three points. Moreover, the 1D calibration object does not have its own occlusion and can be simultaneously observed by all the cameras in a multi-camera system for calibration, and hence the cumbersome and cumulative errors caused by the occlusion in 2D and 3D calibration process can be avoided.

Generally, the 1D calibration algorithm includes two steps: to calculate the closed-form solution of calibration parameters and to optimize the parameter nonlinearly (see the Levenberg–Marquardt (LM) nonlinear nonlinear algorithm in [13,14]). As a typical solution of 1D calibration, Zhang's 1D calibration algorithm [7] is sensitive to the inherent noise on image point and the 1D calibration result is not precise enough. Therefore, the camera parameters initially obtained by the Zhang's 1D calibration algorithm often possess large errors. Franca et al. [15] used the Hartley's normalization algorithm to suppress noise and hence improve the 1D calibration accuracy. In fact, the elements of the 1D calibration measurement matrix are the product of the measurement data, so that each element contains different noise. However, the normalization method does not fully consider the characteristics of each measurement data, so it can only partially attenuate the effects of noise. Kunfeng et al. [16] proposed a similarity-invariant linear based 1D calibration algorithm, which does not require the normalization of image points and has higher precision than the normalized linear algorithm. Moreover, a weighted similarity-invariant linear algorithm (WSILA) is also given in their work, which can improve the calibration accuracy to certain extent. Liang et al. [17] proposed a 1D calibration method for cameras based on Heteroscedastic Error-in-Variables (HEIV) model. In their work, not only the camera parameters but also the image coordinates of the measuring points are optimized. Although the parameter error is further reduced, the computational complexity is high. The above methods have greatly improve the precision of 1D camera calibration when the noise level is low, however when the noise level is increased the calibration precision is decreased dramatically. That is because the rotation angle of the 1D calibration is regarded as a constant value during the nonlinear optimization step, which limits the calibration precision.

In this work, we propose an improved accurate 1D camera calibration based on the WSILA algorithm, which updates the rotation angle of 1D calibration object during the nonlinear optimization step. In this way, the accuracy can be greatly improved. The structure of this paper is organized as follows. The camera model and the 1D calibration object is given in Sect. 2. In Sect. 3, we introduce the principle of the 1D calibration algorithm. In Sect. 4, an improved 1D calibration algorithm is introduced, which combines weighted similar-invariant linear algorithm and an improved nonlinear optimization procedure. Experimental results of both simulation and practical experiments are given in Sect. 5. Finally, conclusions are given in Sect. 6.

2   Preliminaries

2.1   Camera Model

The coordinates of the 3D space point are expressed as $M = [X \; Y \; Z]^T$. The coordinates of the 2D plane point are expressed as $m = [u \; v]^T$. The corresponding homogeneous vector are denoted as $\tilde{M} = [X \; Y \; Z \; 1]^T$ and $\tilde{m} = [u \; v \; 1]^T$, respectively. According to the standard pinhole imaging model, the relationship between point in 3D space and its planar image point can be described as:

$$s\tilde{m} = K\,[R \quad t]\,\tilde{M} \qquad K = \begin{bmatrix} \alpha & \gamma & u_0 \\ 0 & \beta & v_0 \\ 0 & 0 & 1 \end{bmatrix} \qquad (1)$$

where s is the scale factor, R and t are the rotation matrix and the translation vector, respectively, which are used to describe the relationship between the world coordinate system and the camera coordinate system. Without loss of generality, assuming that the camera coordinate system coincides with the world coordinate system, then $R = I_3$ and $t = 0_{1\times3}$, where I_3 is 3×3 the unit matrix. K is the internal parameter matrix of the camera, with α and β are the scale factors in the image u and v axes. γ is the parameter describing the skew of the two image axes, and $[u_0 \; v_0]$ is the coordinates of the principal point. Therefore, Eq. 1 can be expressed as:

$$s\tilde{m} = K\tilde{M} \qquad (2)$$

2.2   1D Calibration Object

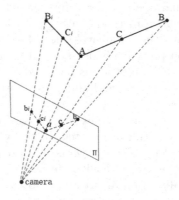

Fig. 1. Illustration of 1D calibration objects.

As shown in Fig. 1, it is assumed that the 1D calibration is composed of three points A, B and C, which satisfies $||C - A|| = L_1$, $||B - A|| = L$. The marker point A is a fixed point, the 1D calibration object rotates around it N times. A, B and C are the coordinates of the 1D calibration at the initial position, and B_i and C_i are the coordinates of points B and C when they are rotated at i-th times. The corresponding image coordinates are a_i, b_i and c_i.

3   The Principle of 1D Calibration

In this section, a variety of calibration algorithms that obtain the camera parameters from multiple observations of an 1D object are detailed.

3.1   Zhang's 1D Calibration Algorithm (ZLA)

According to Fig. 1, since the relative positions of the marker points are known, the collinearity of the three marker points A, B and C can be derived:

$$C = (1 - \lambda) A + \lambda B \tag{3}$$

where $\lambda = L_1/L$, for the convenience of calculation, $\lambda = 0.5$ is usually set. Suppose the projection points of A, B and C are a, b and c, and their Z coordinates are z_A, z_B and z_C, respectively. From Eq. 2, we have:

$$A = z_A K^{-1} \tilde{a}$$
$$B = z_B K^{-1} \tilde{b} \tag{4}$$
$$C = z_C K^{-1} \tilde{c}$$

Substituting the Eq. 4 into the Eq. 3, and simultaneously multiplying on both sides, there is:

$$z_B = -z_A \frac{(1 - \lambda)(\tilde{a} \times \tilde{b})(\tilde{b} \times \tilde{c})}{(\tilde{b} \times \tilde{c})(\tilde{b} \times \tilde{c})} \tag{5}$$

The Euclidean distance between endpoints A and B can be expressed as:

$$L = ||B - A|| = ||z_B K^{-1} \tilde{b} - z_A K^{-1} \tilde{a}|| \tag{6}$$

Substituting them into Eq. 5 yields:

$$z_A \left\| K^{-1} \left(\frac{(1-\lambda)(\tilde{a} \times \tilde{b})(\tilde{b} \times \tilde{c})}{(\tilde{b} \times \tilde{c})(\tilde{b} \times \tilde{c})} \right) + \tilde{a} \right\| = L \tag{7}$$

It is equivalent to:

$$z_A^2 h^T K^{-T} K^{-1} K^{-T} = L^2 \tag{8}$$

$$h = \tilde{a} + \frac{(1 - \lambda)(\tilde{a} \times \tilde{b})(\tilde{b} \times \tilde{c})}{(\tilde{b} \times \tilde{c})(\tilde{b} \times \tilde{c})} + \tilde{b}$$

Let

$$\omega = K^{-T}K^{-1} = \begin{bmatrix} \omega_{11} & \omega_{12} & \omega_{13} \\ \omega_{12} & \omega_{22} & \omega_{23} \\ \omega_{13} & \omega_{23} & \omega_{33} \end{bmatrix}$$

$$= \begin{bmatrix} \frac{1}{\alpha^2} & -\frac{\gamma}{\alpha^2\beta} & \frac{v_0\gamma - u_0\beta}{\alpha^2\beta} \\ -\frac{\gamma}{\alpha^2\beta} & \frac{\gamma^2}{\alpha^2\beta^2} + \frac{1}{\beta^2} & -\frac{\gamma(v_0\gamma - u_0\beta)}{\alpha^2\beta^2} - \frac{v_0}{\beta^2} \\ \frac{v_0\gamma - u_0\beta}{\alpha^2\beta} & -\frac{\gamma(v_0\gamma - u_0\beta)}{\alpha^2\beta^2} - \frac{v_0}{\beta^2} & -\frac{(v_0\gamma - u_0\beta)^2}{\alpha^2\beta^2} + \frac{v_0^2}{\beta^2} + 1 \end{bmatrix} \tag{9}$$

The matrix is a symmetric matrix, which is defined as

$$\varpi = \begin{bmatrix} \omega_{11} & \omega_{12} & \omega_{22} & \omega_{13} & \omega_{23} & \omega_{33} \end{bmatrix}^T \tag{10}$$

and $x = z_A^2$. Let $v = \begin{bmatrix} h_1^2 & 2h_1h_2 & h_2^2 & 2h_1h_3 & 2h_2h_3 & h_3^2 \end{bmatrix}^T$, therefore, Eq. 8 is rewritten as:

$$v^T x = L^2 \tag{11}$$

When observing N images of a 1D object, we get Eq. 12 by stacking n such Eq. 11:

$$\mathbf{V}^T x = L^2 \mathbf{1} \tag{12}$$

where $\mathbf{V} = [v_1, \cdots, v_n]^T$ and $\mathbf{1} = [1, \cdots, 1]^T$. Then x can be solved according to Eq. 13:

$$x = L^2(\mathbf{V}^T\mathbf{V})^{-1}\mathbf{V}^T\mathbf{1} \tag{13}$$

According to x, the symmetric matrix can be acquired. After a simple matrix operation, the parameter K can be finally obtained to realize the camera 1D calibration.

3.2   Franca et al.'s Normalized Linear Algorithm (FNLA)

To reduce the impact of image noise on Zhang's linear algorithm, Franca et al. calibrate the camera using normalized image points. The normalization steps are operated as follows:

(1) *Translating the 2D coordinate point so that its centroid is at the origin;*
(2) *Scale the points so that their average distance to the origin is equal to $\sqrt{2}$;*
(3) *The above transformation is performed independently for each image.*

In this work, the image normalization matrix is denoted as T that is applied to transform the image point $\tilde{m}$ for obtaining the normalized image points $\hat{\tilde{m}} = T\tilde{m}$. In this way, Eq. 2 can be written as $s\hat{\tilde{m}} = TKM$. Let $\hat{K} = TK$, which represents the intrinsic parameters of camera corresponding to the projection $\hat{m}$. Once the internal parameter $\hat{K}$ is estimated in the coordinate system defined by T, the internal parameter k can be obtained by $K = T^{-1}\hat{K}$.

3.3 Nonlinear Optimization

The linear solution of the camera parameters is inaccurate, therefore a non-linearly optimization of the parameters are required. As a classical nonlinear optimization algorithm, the LM algorithm combines the advantages of the steepest descent method and the Gauss-Newton method, and therefore can converge quickly. In 1D calibration, the optimization criterion is normally defined involving the projection of points in the calibration:

$$\sum_{i=1}^{N}(||a_i - a_i'(K, A_i)||^2 + ||b_i - b_i'(K, B_i)||^2 + ||c_i - c_i'(K, C_i)||^2) \qquad (14)$$

where N is the number of captured pattern images. Firstly, as the camera parameters are solved by linear equation written in Eq. 13, the 3D coordinates of the marker points can be recovered according to Eq. 4. Secondly, the rotation angle which is denoted by (θ_i, ϕ_i) can be calculated, which is regarded as a constant that is used to update the 3D coordinates of B_i and C_i based on K and the 3D coordinates of A after each nonlinear optimization iteration. Thirdly, LM algorithm is used to optimize the K and the coordinates of point A in order to minimize the loss written in Eq. 14. In this way, the optimized camera parameters are obtained.

4 Improve Accurate 1D Calibration Algorithm Based on WSILA

4.1 Weighted Similarity-Invariant Linear Algorithm (WSILA)

According to the literature [16], the reciprocal of the standard deviation of the estimated relative depth from different images is used as the weight on the constraint equations in the similar invariant linear calibration algorithm, and a weighted similarity-invariant linear calibration algorithm with higher precision is proposed. According to Eq. 3, we have:

$$Lz_c\tilde{c} = (L - L_1)z_A\tilde{a} + L_1 z_B\tilde{b} \qquad (15)$$

Equation 16 gives three linear equations of z_A, z_B and z_C and dividing both sides of the equal sign by z_A, the equation is equivalent to:

$$L(\tilde{c} - \tilde{b})\eta = (L - L_1)(\tilde{a} - \tilde{c}) \qquad (16)$$

where

$$\eta = \frac{L_1(L - L_1)(\tilde{a} - \tilde{c})^T(\tilde{c} - \tilde{b})}{L_1^2||\tilde{c} - \tilde{b}||^2} \qquad (17)$$

Therefore, the weight ρ of the constraint equation is:

$$\rho = \frac{1}{std(\eta)} \approx \frac{||\tilde{a} - \tilde{b}||}{\eta^2} \qquad (18)$$

Equation 13 is equivalent to:

$$\rho \mathbf{V} x = \rho L^2 \mathbf{1} \tag{19}$$

The solution is then given by:

$$x = \rho L^2 (\rho \mathbf{V} (\rho \mathbf{V})^T)^{-1} (\rho \mathbf{V})^T \tag{20}$$

According to the obtained x, the symmetric matrix $\hat{\varpi}$ is represented. After a simple matrix operation, the parameter K can be finally obtained to realize the camera 1D calibration.

4.2   The Improved Nonlinear Optimization Procedure (INOP)

According to Zhang's 1D calibration algorithm, one can reconstruct the 3D coordinates of Bi after the linear step:

$$B_i = A_i + L \left[\sin \theta_i \cos \phi_i \quad \sin \theta_i \sin \phi_i \quad \cos \theta_i \right]^T \tag{21}$$

And then θ_i and ϕ_i can be calculated as written in Eq. 20:

$$\theta_i = arccos(\frac{z_{B_i} - z_{A_i}}{L}), \quad \phi_i = arccos(\frac{x_{B_i} - x_{A_i}}{L \sin \theta_i}) \tag{22}$$

In the previous algorithm, the 1D calibration object rotation angles θ_i and ϕ_i is considered as a constant during the nonlinear optimization procedure that is calculated according to Eq. 14. Instead, the constant θ_i and ϕ_i are used to update the B_i and C_i according to Eq. 21 that are used to calculated the loss according to Eq. 14. That is inaccurate since the z_{B_i}, z_{A_i}, x_{B_i}, x_{A_i} used in Eq. 22 are inaccurate. Consequently, the accuracy of the nonlinear optimization procedure is limited. To address this limitation, the work propose an improved nonlinear optimization procedure that update the 1D calibration object rotation angle θ_i and ϕ_i by the camera parameters and the coordinates of the rotation point coordinates should after each iteration, as is shown in Algorithm 1. And the objective function at this time is:

$$\sum_{i=1}^{N} \left(||a_i - a_i'(K, A_i, \theta_i, \phi_i)||^2 + ||b_i - b_i'(K, B_i, \theta_i, \phi_i)||^2 + ||c_i - c_i'(K, C_i, \theta_i, \phi_i)||^2 \right) \tag{23}$$

In the actual process, the 1D calibration object rotation angles θ_i and ϕ_i directly affect the 3D coordinates of the endpoints A and B_i, and the 3D coordinates are determined by the camera parameters and the image coordinates, wherein the image coordinates do not change, so when the camera parameters are updated, θ_i and ϕ_i also changes. Therefore, the improved nonlinear optimization procedure used in this paper: 1D calibration object rotation angle θ_i and ϕ_i should be updating by the camera parameters and the coordinates of the rotation point coordinates (show in Algorithm 1). And the objective function at this time is:

Algorithm 1. The improved nonlinear optimization procedure.

Input:

2D coordinates of the marker points $\tilde{m}_{mat}$, 1D calibration object lenth L ($L = ||B_i - A_i||$), λ ($\lambda = L_1/L$, $L_1 = ||C_i - A_i||$, A_i,B_i and C_i are the 3D coordinates of the marker points);

Output:

the camera parameter K

1: {Procedure. $lsqnonlin(\tilde{m}_{mat}, L, \lambda)$}

2: **while** flag is equal to 1 **do**

3:     $[K, z_A, \eta] = WSILA(\tilde{m}_{mat}, L, \lambda)$, $\eta_i = -z_{B_i}/z_A$, z_{B_i} is the depth of the free endpoint B_i, z_A to the depth of the fixed endpoint A and η_i is the relative depth,which is the ratio of z_{B_i} to z_A;

4:     $A_i = z_{A_i} K^{-1} \tilde{a}_i$, $\tilde{a}_i$ is the 2D image ponit of A_i;

5:     $A_{mean} = mean(A_i)$; $z_A = A_{mean}(3)$; $z_{B_i} = -z_A * \eta_i$;

6:     $B_i = z_{B_i} K^{-1} \tilde{b}_i$, $\tilde{b}_i$ is the 2D image point of B_i;

7:     $\theta_i = arccos(\dfrac{z_{B_i} - z_A}{L})$, $\phi_i = arccos(\dfrac{x_{B_i} - x_A}{L * sin\theta_i})$, θ_i,ϕ_i are rotation angle of 1D calibration objects;

8:     $\tilde{m}_{cal} = 3Dto2D(A, B)$, according to the standard pinhole imaging model, $\tilde{m}_{cal}$ is the updated 2D coordinates of the marker points;

9:     Set the objective function F to $\sum_{i=1}^{N} (||a_i - a'_i(K, A_i, \theta_i, \phi_i)||^2 + ||b_i - b'_i(K, B_i, \theta_i, \phi_i)||^2 + ||c_i - c'_i(K, C_i, \theta_i, \phi_i)||^2)$;

10:     Optimizing camera parameters K and the 3D coordinates of the fixed endpoint A using least squares, and update θ_i,ϕ_i;

11:     **if** F reach the minimum value using the least squares **then**

12:         $output(K)$;

13:         $flag = 0$;

14:     **else**

15:         $\tilde{m}_{mat} = \tilde{m}_{cal}$;

16:         $flag = 1$;

17:     **end if**

18: **end while**

---

5 Experimental Results

In this section, both simulation and laboratory experiments are conducted to test the performance of improved nonlinear optimization procedure. In detail, the 1D calibration accuracy of six algorithms are tested, including (a) Zhang's linear algorithm that is referred as ZLA; (b) Zhang's linear algorithm combined with the improved nonlinear optimization procedure that is referred as ZLA+INOP; (c) Franca et al.'s normalized linear algorithm that is referred as FNLA; (d) Franca et al.'s normalized linear algorithm combined with the improved nonlinear optimization procedure that referred as FNLA+INOP; (e) Weighted similarity-invariant linear algorithm that is referred as WSILA; (f) Weighted similarity-invariant linear algorithm combined with the improved nonlinear optimization procedure that is referred as WSILA+INOP.

5.1   Simulations

The pinhole camera model and the 1D camera calibration process are simulated in Matlab. The parameters of the simulated camera are set as follows: $\alpha = 1000$, $\beta = 1000$, $\gamma = 0$, $u_0 = 320$ and $v_0 = 240$. The image resolution is 640×80 pixels. A stick with a length of $70\,\mathrm{cm}$ is simulated, in which the 3D coordinates of the fixed point A is $\begin{bmatrix} 0 & 40 & 150 \end{bmatrix}^T$, the other endpoint of the stick is B, while C is located in the middle of A and B. This experiment generates sticks with multiple random directions by sampling $\theta_i = [\pi/6, 5\pi/6]$, $\phi_i = [\pi/6, 5\pi/6]$ according to uniform distribution. The corresponding image coordinates a_i, b_i, c_i can be acquired through projecting A_i, B_i and C_i onto the image via pinhole camera model. Gaussian noise with a mean of 0 and different standard deviations (noise level) is added to the projected image points a, b and c, based on which the camera parameters are estimated using the six camera calibration algorithms separately. Then the relative errors of calibrated parameters are calculated, as

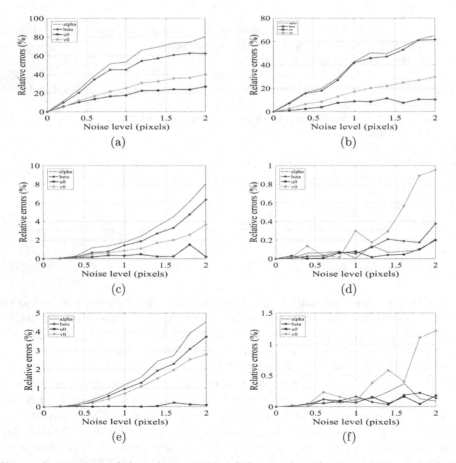

Fig. 2. Comparison of the relative errors of the six algorithms under the noise level from 0 to 2 pixels: (a) ZLA, (b) ZLA+INOP, (c) FNLA, (d) FNLA+INOP, (e) WSILA, and (f) WSILA+INOP.

shown in Fig. 2. In the experiment, the noise level was increased from 0 pixels to 2 pixels in a step of 0.2 pixels. For each noise level, 150 independent experiments were performed, and the average of each parameter was calculated.

As shown in Fig. 2, compared to ZLA (Fig. 2(a)), both FNLA (Fig. 2(c)) and WSILA (Fig. 2(e)) can greatly reduce the relative errors of calibrated parameters. Since it is taken into consideration of the different levels of importance of the constraint equations constructed from different poses, WSILA can provide better performance than FNLA. However, since the rotation angles θ_i and ϕ_i are not updated, the accuracy of WSILA is still limited, which is improved by the proposed nonlinear optimization procedure. When comparing the results of three combinations of (a) and (b), (c) and (d), (e) and (f), it is clear that the proposed nonlinear optimization procedure can greatly reduce the relative errors. For example, when the noise level is 2 pixels, the relative errors of camera parameters obtained by the combination of WSILA and the proposed nonlinear optimization procedure are smaller than 1.2%, while most of them are smaller than 0.2%. And the optimization process takes an average of 0.0078s to converge. Compared with the original WSILA, the average time calculated by WSILA+INOP algorithm increases from 1.0039s to 0.9941s. Therefore, the proposed nonlinear optimization procedure can greatly improve the camera calibration accuracy.

5.2 Laboratory Experiments

Fig. 3. Sample images of a 1D object used for camera calibration.

To verify the performance of proposed nonlinear optimization procedure, laboratory experiments are conducted. Three table tennis balls were fixed on a stick. The distance between adjacent balls are 133 cm (that is, L = 266 cm). One of the endpoints is fixed and the stick rotated 150 times around the fix endpoint, some of which are shown in Fig. 3. Then, the table tennis balls are detected through Hough circle detection algorithm, and the image coordinates of the marker points are measured accordingly. Then, the camera calibration is performed using the six algorithms. In order to evaluate the performance of the algorithms, the camera parameters are calibrated by Zhang's 2D camera calibration as a baseline, as is shown in Fig. 4. The experimental results are shown in Table 1.

From Table 1, the results of the WSILA combined with the nonlinear optimization algorithm (the proposed nonlinear optimization procedure) are closer to the Zhang's 2D calibration results compared to the other five methods. That is consistent with the simulation results.

Fig. 4. Sample image of the planar pattern used for camera calibration.

Table 1. 1D calibration results of cameras with different algorithms.

Solution	α	β	γ	u_0	v_0
ZLA	1230	1213	27.4	915	517
FNLA	1714	1717	22.5	1008	683
WSILA	1230	1213	27.4	915	517
ZLA+INOP	1339	1238	19.0	1002	544
FNLA+INOP	1750	1749	16.3	1004	689
WSILA+INOP	1747	1745	9.8	999	692
Zhang's 2D Camera Calibration	1742	1739	5.0	1005	754

6   Conclusion

Despite the flexibility of the Zhang's 1D calibration algorithm, the calibration results have large errors. To solve this problem, several solutions including simple data normalization and weighted similar invariant linear algorithms have been proposed to improve the calibration results. In this work, an improved nonlinear optimization procedure is proposed to further reduces the relative error of the camera parameters calibration results.

Through simulation experiments and laboratory experiments, the performance of the six 1D camera algorithms are studied. The results show that the method combining the weighted similarity invariant linear algorithm and the improved nonlinear optimization procedure has the best performance, which is better than the combination of the Zhang's linear algorithm and the improved nonlinear optimization procedure. It is also superior to the combination of the normalized and the improved nonlinear optimization procedure. According to the simulation results, when the noise level is 2 pixels, the relative errors of camera parameters obtained by the combination of WSILA and the improved nonlinear optimization procedure are smaller than 1.2%, while most of them are smaller than 0.2%. Through laboratory experiments, the parameters calibrated through the combination of WSILA and the improved nonlinear optimization procedure is the closest ones to the results from Zhang's 2D camera calibration. Therefore, the proposed nonlinear optimization procedure can greatly improve the camera calibration accuracy.

Acknowledgements. This work is financially supported in parts by the Fujian Provincial Department of Science and Technology of China (Grant No. 2019H0006 and 2018J01774), the National Natural Science Foundation of China (Grant No. 61601127), and the Foundation of Fujian Provincial Department of Industry and Information Technology of China (Grant No. 82318075).

References

1. Tsai, R., et al.: A versatile camera calibration technique for high-accuracy 3D machine vision metrology using off-the-shelf TV cameras and lenses. IEEE J. Rob. Autom. **3**(4), 323–344 (2003)
2. Yuan, T., Zhang, F., Tao, X.P.: Flexible geometrical calibration for fringe-reflection optical three-dimensional shape measurement. Appl. Opt. **54**(31), 9102–9107 (2015)
3. Yoo, J.S., et al.: Improved LiDAR-camera calibration using marker detection based on 3D plane extraction **13**(6), 2530–2544 (2018)
4. Duan, F., Wu, F., et al.: Pose determination and plane measurement using a trapezium. Pattern Recogn. Lett. **29**(3), 223–231 (2008)
5. Li, N., Hu, Z.Z., Zhao, B.: Flexible extrinsic calibration of a camera and a two-dimensional laser rangefinder with a folding pattern. Appl. Opt. **55**(9), 2270–2280 (2016)
6. Zhang, Z.Y.: Flexible camera calibration by viewing a plane from unknown orientation. In: IEEE International Conference on Computer Vision (1999)
7. Peng, E., Li, L.: Camera calibration using one-dimensional information and its applications in both controlled and uncontrolled environments. Pattern Recogn. **43**(1), 1188–1198 (2010)
8. Zhang, Z.Y.: Camera calibration with one-dimensional objects. IEEE Trans. Pattern Anal. Mach. Intell. **26**(7), 892–899 (2004)
9. Wang, L., Wang, W.W., et al.: A convex relaxation optimization algorithm for multi-camera calibration with 1D objects. Neurocomputing **215**, 82–89 (2016)
10. Lv, Y.W., Liu, W., et al.: Methods based on 1D homography for camera calibration with 1D objects. Appl. Opt. **57**(9), 2155–2164 (2018)
11. Akkad, N.E., Merras, M., et al.: Camera self-calibration with varying intrinsic parameters by an unknown three-dimensional scene. Int. J. Comput. Graphics. **30**(5), 519–530 (2014)
12. Sun, Q., Wang, X., et al.: Camera self-calibration with lens distortion. Optik Int. J. Light Electron Optics. **127**(10), 4506–4513 (2016)
13. Finsterle, S., Kowalsky, M.B.: A truncated Levenberg-Marquardt algorithm for the calibration of highly parameterized nonlinear models **37**(5), 731–738 (2011)
14. Li, J.H., Zheng, W.X., et al.: The Parameter estimation algorithms for Hammerstein output error systems using Levenberg-Marquardt optimization method with varying interval measurements. IFAC Papersonline **48**(8), 457–462 (2015)
15. Franca, J.D., Stemmer, M.R., et al.: Revisiting Zhang's 1D calibration algorithm. Pattern Recogn. **43**(3), 1180–1187 (2010)
16. Kunfeng, S., Qiulei, D., Fuchao, W.: Weighted similarity-invariant linear algorithm for camera calibration with rotating 1D objects. IEEE Trans. Image Process. **21**(8), 3806–3812 (2012)
17. Wang, L., Duan, F.: Zhang's one-dimensional calibration revisited with the heteroscedastic error-in-variables model. In: 8th IEEE International Conference on Image Processing. IEEE (2011)

AAT: An Efficient Model for Synthesizing Long Sequences on a Small Dataset

Quan Anh Minh Nguyen, Quang Trinh Le[✉], Huu Quoc Van,
and Duc Dung Nguyen

HO CHI MINH City University of Technology, Ho Chi Minh City, Vietnam
nguyenquananhminh@gmail.com, trinhle.cse@gmail.com, rin2401@gmail.com,
nddung@hcmut.edu.vn

Abstract. This work represents an alternative model for speech synthesis, which addresses some major disadvantages of current end-to-end models. Current state-of-the-art models still have some troubles while dealing with long sentences and the size of the dataset. Our proposed Adaptive Alignment Tacotron (AAT) model, however, has achieved impressive results for a very small dataset of the Vietnamese language. By leveraging the nature of diagonal alignment between phoneme and acoustic sequences, we address the issue with long sequences and the proposed model can also be trained efficiently on the small dataset (5 h of recording). The proposed model consists of the following components: a stacked convolutional encoder, a local diagonal attention module, a decoder with schedule teacher forcing to produce a coarse mel-spectrogram prediction, and a converter to transform the mel-spectrogram to linear-spectrogram. Experimental results show that the proposed model achieves faster convergence speed and higher stability than the baseline model and open a feasible approach for speech synthesis on languages with small dataset.

Keywords: Text-to-speech · Attention · Encoder-decoder ·
Sequence-to-sequence · Tacotron · Vietnamese speech synthesis

1 Introduction

The Text-to-speech (TTS) systems play an important role in a variety of applications, such as audio-book, human-computer interaction, media and entertainment. The quality of a TTS system depends primarily on two main factors: its **intelligibility** and **naturalness**. Intelligibility specifies the clarity of synthesized audio, while naturalness describes information about ease of listening, global stylistic consistency, regional or language level nuances, among others.

Traditional TTS systems typically consist of many domain-specific modules [28]. For example, a typical parametric TTS system [29,35] is an integration of many modules such as text analyzer, F0 generator, duration predictor, spectrum

© Springer Nature Switzerland AG 2019
R. Chamchong and K. W. Wong (Eds.): MIWAI 2019, LNAI 11909, pp. 76–86, 2019.
https://doi.org/10.1007/978-3-030-33709-4_7

generator and a vocoder that synthesize a waveform from these data. However, the audio produced by these systems often unnatural compared to human speech.

Current state-of-the-art TTS systems are mostly based on deep neural models, which are all data-hungry. Although deep neural network systems are sometimes criticized as a black box compared the traditional TTS systems, they have obtained incredible results. An end-to-end TTS system named Tacotron [22, 33] can directly produce spectrogram feature from input text without any domain-specific knowledge utilize the sequence-to-sequence with attention framework [2, 4, 26]. One shortcoming of these end-to-end systems is that they require large amounts of data to training effectively and can't deal with long sentences, especially the sentences are longer than the longest sentence in the training set, which brings challenges on real-world problems and many languages that are scarce of paired speech and text data like Vietnamese.

In this work, we propose an architecture for Vietnamese speech synthesis named Adaptive Alignment Tacotron (AAT), address several real-world issues that arise when attempting to deploy an attention-based TTS system. Specifically, we make the following contributions:

- We propose an end-to-end character-to-spectrogram architecture, which leveraging parallel computation from convolution network to train faster than architectures using the recurrent neural network (e.g. [22, 33]).
- We introduce the local diagonal attention for our end-to-end speech synthesis model to address issues of long sentences.
- We show that our architecture can be trained efficiently on small Vietnamese dataset (5 h) with higher stability than the baseline model [22].
- Our AAT model is the first end-to-end attention-based seq2seq for Vietnamese TTS system.

2   Related Work

2.1   Deep Learning for TTS

To our best knowledge, the first end-to-end speech synthesis model is [32]. However, this model needs the pretrained hidden Markov model (HMM) aligner to help sequence-to-sequence model learning alignment and some technique for training model efficiently. Char2Wav [24] describes yet another similar approach but this model still has to predict complex characteristics of sound before using SampleRNN [13] to synthesized sound based on those characteristics. In the model of Deep Voice 1 [1] and Deep Voice 2 [8], components of traditional TTS system (grapheme-to-phoneme conversion, duration-frequency prediction model and waveform synthesis) were replaced with the corresponding neural network. However, each component is independently trained and it is difficult to end-to-end training. Deep Voice 3 [18] employs an attention-based sequence-to-sequence model, yielding a more compact architecture. However, its naturalness has not been shown to rival that of human speech. Wavenet [16] is the kind of a Vocoder, which synthesizes a waveform from some conditioning information (e.g spectrogram, F0, duration) and achieved state-of-the-art for English Text-to-speech system. The most extreme model in speech synthesis is Tacotron [22,33], which depends barely on mel-spectrograms and linear spectrograms, and not on any other complex speech features e.g. F0, duration. In Tacotron 2, it uses Wavenet as vocoder instead Griffin-Lim [10] in Tacotron. However, both Tacotron and Tacotron 2 has a drawback that it can't deal with long sentences and costly to train due to using the recurrent neural network (RNN) throughout the model. Some studies try to solve those problem by enforces monotonic attention in Deep Voice 3 [18], Gaussian mixture model (GMM) attention [9] in Char2wav [24], guild attention in Deep convolution TTS [27] and use fully-convolutional character-to-spectrogram architecture to speed up training [18,27]. However, none of those methods as good as Tacotron in terms of naturalness. Our model rather than a spectrogram prediction stage, and use simple vocoder Griffin-Lim [10] for simplicity.

2.2   Sequence to Sequence (seq2seq) Learning

Sequence-to-sequence models [26] encode a variable-length input into hidden states, which are then processed by an RNN-based decoder to produce a target sequence. An RNN-based seq2seq, however, has disadvantages that a vanilla encoder-decoder model can't encode too long sequence into a fixed-length vector effectively. This problem has been resolved by the attention mechanism [2]. Attention-based seq2seq are widely applied in machine translation [2,31], speech recognition [4], text summarization [14,21], and speech synthesis [18,22,27]. Model Tacotron++ demonstrates the utility of diagonal attention during training in TTS and builds upon Tacotron [22,33], but higher stability than Tacotron and can be synthesized long sentences, makes attention-based TTS feasible for a TTS system.

2.3   Vietnamese Text-to-speech System

Several recent works tackle the problem of synthesizing Vietnamese speech, include HMM-based speech synthesis [15,17], non-uniform unit selection [6]. The first deep neural network for Vietnamese TTS system in our best knowledge is [30]. However, all of these approaches include multiple steps, extracting complex audio features (e.g F0, duration, spectral envelope, aperiodicities). Our AAT model is the first end-to-end attention-based seq2seq for Vietnamese TTS system, which synthesizes audio conditioned only spectrogram feature.

3   Proposed Appproach

Our proposed model inherits from Tacotron 2 [22] composes of two main components: spectrogram predicted model, which based on sequence-2-sequence architecture incorporating with attention mechanism and Converter convert the mel-spectrogram into linear-spectrogram for waveform synthesis. First, we use a faster and lighter encoder than encoder of Tacotron 2. Second, we propose a new attention mechanism to overcome the long sentence synthesis problem. Third, we apply multi-head attention architecture [31] to convert from mel-spectrogram into linear spectrogram serving synthesize better audio wave. Figure 1 depicts the overview of our model.

3.1   Convolution Encoder

The encoder transforms a sequence of characters into a hidden representation, which serves decoder predicts spectrogram. Each input character is present as learnable embedding 512-d vector. Instead of passing character embedding through three convolution layer and one Bi-LSTM layer like Tacotron 2 [22], character embedding go through a sequence of special convolution block shown in Fig. 1. The convolution block consists of 1-D convolution with filter size of five and a gated linear unit [5] with a residual connection. Gated linear unit is used to alleviate the vanishing gradient issue for stacked convolution blocks while retaining non-linearity. In this work, out encoder composes of 3 convolution block. Outside of these blocks, we add one more residual connection link

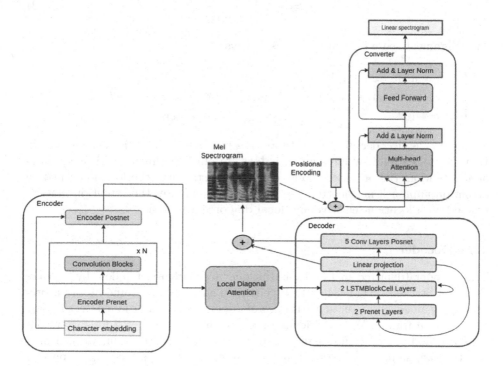

Fig. 1. AAT uses stacked Dilated Convolution Blocks to encode character input into encoder hidden state for an attention-based decoder. The output of the decoder is mel-spectrogram then feed into Multi-head attention for producing linear-spectrogram for waveform synthesis by Griffin-Lim [10] vocoder.

between character embedding and output of convolution blocks. This connection gives hidden encoder vectors more information about the character and lets convolution blocks focus on learning context around the character.

3.2   Local Diagonal Attention

Our local diagonal attention is the extend version of location-sensitive attention [4] in Tacotron 2 [22]. The motivation of the local diagonal attention is that the output decoder frame is conditioned on the part of whole encoder hidden state and move forward with decoder timesteps. The Attention's task is producing attention context vector c conditioned on the encoder hidden state $(h_j, ..., h_{T_x})$ at decoder timesteps i, where T_x is the length of input character sentence.

$$c_i = \sum_{j=1}^{T_x} \alpha_{ij} h_j \tag{1}$$

The weight α_{ij} of each h_j is calculated by the formula:

$$\alpha_{ij} = \frac{exp(e_{ij})}{\sum_{k=1}^{T_x} exp(e_{ik})} \tag{2}$$

The alignment score e_{ij} is modeled as follows:

$$\begin{cases} k_{i-1} = argmax(\alpha_{i-1,:}) \\ e_{ij} = w^T tanh(W s_{i-1} + V h_j + U \alpha_{i-1,j} + b) & j \in [k_{i-1} - \beta, k_{i-1} + \beta] \\ e_{ij} = -10^6 & other \end{cases} \tag{3}$$

where s_{i-1} is decoder hidden state at timestep $i-1$ and β is window width. W, V, U are learnable weights and b is bias. In our experiment, with local diagonal attention, the proposed model converges faster than the location-sensitive attention approach in Tacotron 2 [22]. Figure 2 compares the attention matrix, trained with location-sensitive and local diagonal attention.

3.3   Decoder

The decoder (depicted in Fig. 1), is the one decoder prenet follow by 2-layers Residual Stacked LSTM [34], which generates spectrogram frame in an autoregressive manner by predicting r-frames of spectrogram conditioned on the past spectrogram frames. We empirically observed and choose $r = 2$ for faster convergence and better audio quality than $r = 1$. For spectrogram, we use mel-spectrogram similar to Tacotron [33] as the compact low-dimensional spectrogram frame representation.

The decoder starts with a GO frame which is all zero. Every time-step, the previous time frame is fed as the input of the decoder prenet. The decoder *pre-net* containing 2 fully connected layers of 256 hidden Swish [19] units. The

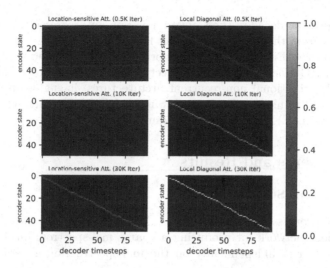

Fig. 2. Comparison of the attention matrix, trained with location-sensitive and local diagonal attention. (Left) is location-sensitive attention and (Right) is local diagonal attention.

prenet output and attention context vector are concatenated and passed through 2-layers Resudual Stacked uni-directional LSTM layers with 1024 units. The output of LSTM is concatenated with attention context vector then projected through a linear transform to predict the target mel-spectrogram frame and scalar *stop token*. This "stop token" prediction is used during inference to allow the model to dynamically determine when to stop the decoding process. Specifically, the decoding process completes at the first frame for which the output sigmoid of "stop token" exceeds a threshold of 0.5. Finally, the predicted mel-spectrogram is passed through a 3-layer convolutional *post-net* which predicts a residual to add to the mel-spectrogram prediction to improve the overall reconstruction. Each post-net layer is comprised of 384 filters with kernel size is 5 with batch re-normalization [11], followed by tanh activations except the final layer. We empirically observed and use the mean absolute error (MAE) on the mel-spectrogram before and after the post-net to faster convergence.

For regularization, we use Drop-block [7] with keep-probability 0.8 for all convolution layers in post-net and Zone-out [12] with keep-probability 0.85 for LSTM layers. Drop-out [25] is applied on pre-net layers with probability 0.5 for both training and inference.

3.4 Converter

The Converter transforms the mel-spectrogram output from the Decoder to a target that can be synthesized into waveforms. Since we use Griffin-Lim [10] as the vocoder, the Converter learns to predict linear-scale log-magnitude spectrogram. Unlike the decoder, the Converter is non-autoregressive, so it can use future context from the decoder to predict its outputs.

We employ the Encoder from Transformer [31] as Converter (see Fig. 1). Unlike the CBHG block consists of bi-directional RNN on Tacotron [33], Transformer is fully-attentional non-recurrent architectures, it enables fully parallel computation on training and inference. Similar to the Decoder, we use mean absolute error (MAE) for linear-spectrogram prediction.

4 Experiments and Results

4.1 Experiment Setup

Our training process involves two steps. We first train the attention-based encoder-decoder network for producing a coarse mel-spectrogram, followed by training a Converter independently on the mel-spectrogram generated by the first network.

To train the mel-spectrogram prediction network, we apply scheduled teacher-forcing [3] with a batch size of 32 on single GPU 1080. We use AMSGrad [20] and cyclic learning rate [23] for optimization. The β window parameter is 10 for training and 4 for inference to enforces nearly monotonic alignment. Similar to Tacotron 2 [22], we apply L2 regularization with weight 10^{-6}. For faster convergence, we initialize the weight of the encoder by pretrained Vietnamese tone prediction task.

We then train Converter network on mel-spectrogram predicted by the first network. We also train Converter network with Ground-truth mel-spectrogram and use it as pretrained weights. For Converter, batch size of 64 and fixed learning rate of 0.0001 is used.

We train all models on an internal Vietnamese dataset, which contains $\sim$5 h of speech from a single female speaker, $\sim$5K pairs of text&speech. All text in our datasets is spelled out. Find out more sample at[1]

4.2 Evaluation

Training Time: To fairly compared training iteration time, we just train the mel-spectrogram prediction network with same experiment setup above. For our AAT model, the average training time is 0.8 s using one 1080 GPU compared 1.3 s for Tacotron 2, indicating 1.5x faster in training speed. In addition, AAT converges after $\sim$40K iterations while Tacotron 2 requires $\sim$70K iterations. This speedup is due to the convolutional in Encoder and the attention mechanism.

Naturalness: For evaluating the naturalness, we randomly selected 15 samples from the test set of our internal dataset as the evaluation set. Each sample includes the Ground-truth, the Tacotron synthesized and AAT synthesized

---

[1] AAT: The Local Diagonal Attention Approach For End-To-End Speech Synthesis.

audios. Therefore, we have 75 clips for human rating service system. The rater evaluated these 75 clips, rating them from 1 (Bad) to 5 (Excellent). Each rater needs to rate all 75 clips. Table 1 compares the performance of our model (AAT) and the Tacotron as well as ground-truth. Our MOS (95% confidence interval) was 3.65 ± 0.55 while the Tacotron's was 3.52 ± 0.67. Both of them has MOS comparable with ground-truth 3.83 ± 0.43. Although it is not a strict comparison since the experiment setup is difference and Tacotron is re-implementation, it would be still concluded that our model works well on a small dataset and can be compared with Tacotron in terms of naturalness.

Table 1. Comparison of MOS (95% confidence interval), training time, and iterations of an Tacotron and the proposed model (AAT). All models use Griffin-Lim [10] as their Vocoder.

Method	Iteration	Time	MOS (95% CI)
Ground-truth (GT)			3.83 ± 0.43
Tacotron	71K	~1 days	3.52 ± 0.67
AAT (Ours)	39K	~8,6 h	3.65 ± 0.55

Intelligibility: We also evaluate the intelligibility of our model and Tacotron. To do that, we construct 109 sentences that include particularly-challenging cases and then count attention error modes. One sentence is judged to be failed if they are some repeated words, mispronunciations, skipped words and unclear words. Attention error counts are listed in Table 2 and indicate that our model, trained with local diagonal attention, largely outperforms Tacotron use location-sensitive attention.

Table 2. Attention error counts for Tacotron and AAT on the 109 sentences test set. One sentence is judged to be failed if they are some repeated words, mispronunciations, and skipped words.

Method	Failed	Complete
Tacotron	57	52
AAT (Ours)	21	88

Synthesized Long Sentences: To investigate the impact of our proposed attention mechanism, we conduct synthetic experiments on 50 long sentences from 30 s to 15 min. In our experiments, Tacotron 2 failed all 50 sentences while our model, AAT, complete all sentences. This experiment confirms that our model is suitable for production.

5   Conclusions

The paper describes a first end-to-end Attention-based neural network for Vietnamese TTS system. We proposed a new approach that can address issues of previous models and make a significant improvement in the synthesizing process. As a result, our propsed model can handle long sequences, and almost any length sequences by far. In addition, our proposed approach works well under small dataset. The results show that AAT can synthesize high-quality speech as good as the groundtruth with respect a very small dataset. This indicate a huge potential for real-applications where the data resource is limited.

References

1. Arık, S.Ö., et al.: Deep voice: Real-time neural text-to-speech. In: Precup, D., Teh, Y.W. (eds.) Proceedings of the 34th International Conference on Machine Learning. Proceedings of Machine Learning Research, vol. 70, pp. 195–204. PMLR, International Convention Centre, Sydney (2017). http://proceedings.mlr.press/v70/arik17a.html
2. Bahdanau, D., Cho, K., Bengio, Y.: Neural machine translation by jointly learning to align and translate. CoRR abs/ 1409.0473 (2014)
3. Bengio, S., Vinyals, O., Jaitly, N., Shazeer, N.: Scheduled sampling for sequence prediction with recurrent neural networks. In: Cortes, C., Lawrence, N.D., Lee, D.D., Sugiyama, M., Garnett, R. (eds.) Advances in Neural Information Processing Systems, vol. 28, pp. 1171–1179. Curran Associates, Inc. (2015). http://papers.nips.cc/paper/5956-scheduled-sampling-for-sequence-prediction-with-recurrent-neural-networks.pdf
4. Chorowski, J.K., Bahdanau, D., Serdyuk, D., Cho, K., Bengio, Y.: Attention-based models for speech recognition. In: Cortes, C., Lawrence, N.D., Lee, D.D., Sugiyama, M., Garnett, R. (eds.) Advances in Neural Information Processing Systems, vol. 28, pp. 577–585. Curran Associates, Inc. (2015). http://papers.nips.cc/paper/5847-attention-based-models-for-speech-recognition.pdf
5. Dauphin, Y.N., Fan, A., Auli, M., Grangier, D.: Language modeling with gated convolutional networks. CoRR abs/1612.08083 (2016). http://arxiv.org/abs/1612.08083
6. Do, T.V., Tran, D.D., Nguyen, T.T.T.: Non-uniform unit selection in Vietnamese speech synthesis. In: SoICT (2011)
7. Ghiasi, G., Lin, T., Le, Q.V.: DropBlock: a regularization method for convolutional networks. In: Bengio, S., Wallach, H.M., Larochelle, H., Grauman, K., Cesa-Bianchi, N., Garnett, R. (eds.) Advances in Neural Information Processing Systems 31: Annual Conference on Neural Information Processing Systems 2018, NeurIPS 2018, Montréal, Canada, 3–8 December 2018, pp. 10750–10760 (2018). http://papers.nips.cc/paper/8271-dropblock-a-regularization-method-for-convolutional-networks
8. Gibiansky, A., et al.: Deep voice 2: multi-speaker neural text-to-speech. In: Guyon, I., et al. (eds.) Advances in Neural Information Processing Systems 30, pp. 2962–2970. Curran Associates, Inc. (2017). http://papers.nips.cc/paper/6889-deep-voice-2-multi-speaker-neural-text-to-speech.pdf
9. Graves, A.: Generating sequences with recurrent neural networks. CoRR abs/ 1308.0850 (2013)

10. Griffin, D.W., Lim, J.S.: Signal estimation from modified short-time Fourier transform. In: IEEE International Conference on Acoustics, Speech, and Signal Processing, ICASSP 1983, Boston, Massachusetts, USA, 14–16 April 1983, pp. 804–807. IEEE (1983). https://doi.org/10.1109/ICASSP.1983.1172092
11. Ioffe, S.: Batch renormalization: towards reducing minibatch dependence in batch-normalized models. In: Guyon, I., et al. (eds.) Advances in Neural Information Processing Systems 30: Annual Conference on Neural Information Processing Systems 2017, Long Beach, CA, USA, 4–9 December 2017, pp. 1942–1950 (2017). http://papers.nips.cc/paper/6790-batch-renormalization-towards-reducing-minibatch-dependence-in-batch-normalized-models
12. Krueger, D., et al.: Zoneout: Regularizing RNNs by randomly preserving hidden activations. arXiv e-prints abs/ 1606.01305. https://arxiv.org/abs/1606.01305, June 2016
13. Mehri, S., et al.: SampleRNN: an unconditional end-to-end neural audio generation model. CoRR abs/ 1612.07837 (2016). http://arxiv.org/abs/1612.07837
14. Nallapati, R., Zhou, B., dos Santos, C.N., Gülçehre, Ç., Xiang, B.: Abstractive text summarization using sequence-to-sequence RNNs and beyond. In: Goldberg, Y., Riezler, S. (eds.) Proceedings of the 20th SIGNLL Conference on Computational Natural Language Learning, CoNLL 2016, Berlin, Germany, 11–12 August 2016, pp. 280–290. ACL (2016). http://aclweb.org/anthology/K/K16/K16-1028.pdf
15. Nguyen, T.T.T.: HMM-based vietnamese text-to-speech : prosodic phrasing modeling, corpus design system design, and evaluation. (Text-To-Speech à base de HMM (Hidden Markov Model) pour le vietnamien : modélisation de la segmentation prosodique, la conception du corpus, la conception du système, et l'évaluation perceptive). Ph.D. thesis, University of Paris-Sud, Orsay, France (2015). https://tel.archives-ouvertes.fr/tel-01260884
16. van den Oord, A., et al.: WaveNet: a generative model for raw audio. CoRR abs/ 1609.03499 (2016). http://arxiv.org/abs/1609.03499
17. Phan, T.S., Dinh, A.T., Vu, T.T., Luong, C.M.: An improvement of prosodic characteristics in vietnamese text to speech system. In: Huynh, V., Denoeux, T., Tran, D., Le, A., Pham, S. (eds.) Knowledge and Systems Engineering. Advances in Intelligent Systems and Computing, vol. 244, pp. 99–111. Springer, Cham (2014). https://doi.org/10.1007/978-3-319-02741-8_10
18. Ping, W., et al.: Deep voice 3: scaling text-to-speech with convolutional sequence learning. In: International Conference on Learning Representations (2018). https://openreview.net/forum?id=HJtEm4p6Z
19. Ramachandran, P., Zoph, B., Le, Q.V.: Searching for activation functions. In: 6th International Conference on Learning Representations, ICLR 2018, Vancouver, BC, Canada, 30 April–3 May 2018, Workshop Track Proceedings. OpenReview.net (2018). https://openreview.net/forum?id=Hkuq2EkPf
20. Reddi, S.J., Kale, S., Kumar, S.: On the convergence of adam and beyond. In: 6th International Conference on Learning Representations, ICLR 2018, Vancouver, BC, Canada, 30 April–3 May 2018, Conference Track Proceedings. OpenReview.net (2018). https://openreview.net/forum?id=ryQu7f-RZ
21. Rush, A.M., Chopra, S., Weston, J.: A neural attention model for abstractive sentence summarization. In: Màrquez, L., Callison-Burch, C., Su, J., Pighin, D., Marton, Y. (eds.) Proceedings of the 2015 Conference on Empirical Methods in Natural Language Processing, EMNLP 2015, Lisbon, Portugal, 17–21 September 2015, pp. 379–389. The Association for Computational Linguistics (2015). http://aclweb.org/anthology/D/D15/D15-1044.pdf

22. Shen, J., et al.: Natural TTS synthesis by conditioning WaveNet on Mel spectrogram predictions. CoRR abs/ 1712.05884 (2017). http://arxiv.org/abs/1712.05884
23. Smith, L.N.: Cyclical learning rates for training neural networks. In: 2017 IEEE Winter Conference on Applications of Computer Vision, WACV 2017, Santa Rosa, CA, USA, 24–31 March 2017, pp. 464–472. IEEE Computer Society (2017). https://doi.org/10.1109/WACV.2017.58
24. Sotelo, J., et al.: Char2wav: end-to-end speech synthesis. In: 5th International Conference on Learning Representations, ICLR 2017, Toulon, France, 24–26 April 2017, Workshop Track Proceedings. OpenReview.net (2017). https://openreview.net/forum?id=B1VWyySKx
25. Srivastava, N., Hinton, G., Krizhevsky, A., Sutskever, I., Salakhutdinov, R.: Dropout: a simple way to prevent neural networks from overfitting. J. Mach. Learn. Res. **15**, 1929–1958 (2014). http://jmlr.org/papers/v15/srivastava14a.html
26. Sutskever, I., Vinyals, O., Le, Q.V.: Sequence to sequence learning with neural networks. In: NIPS, pp. 3104–3112 (2014)
27. Tachibana, H., Uenoyama, K., Aihara, S.: Efficiently trainable text-to-speech system based on deep convolutional networks with guided attention. In: 2018 IEEE International Conference on Acoustics, Speech and Signal Processing, ICASSP 2018, Calgary, AB, Canada, 15–20 April 2018, pp. 4784–4788. IEEE (2018). https://doi.org/10.1109/ICASSP.2018.8461829
28. Taylor, P.: Text-to-Speech Synthesis. Cambridge University Press (2009). https://doi.org/10.1017/CBO9780511816338
29. Tokuda, K., Yoshimura, T., Masuko, T., Kobayashi, T., Kitamura, T.: Speech parameter generation algorithms for hmm-based speech synthesis. In: IEEE International Conference on Acoustics, Speech, and Signal Processing. ICASSP 2000, Hilton Hotel and Convention Center, Istanbul, Turkey, 5–9 June 2000, pp. 1315–1318. IEEE (2000). https://doi.org/10.1109/ICASSP.2000.861820
30. Van Nguyen, T., Quoc Nguyen, B., Huy Phan, K., Van Do, H.: Development of Vietnamese speech synthesis system using deep neural networks. J. Comput. Sci. Cybern. **34**, 349–363 (2019). https://doi.org/10.15625/1813-9663/34/4/13172
31. Vaswani, A., et al.: Attention is all you need. In: Guyon, I., et al. (eds.) Advances in Neural Information Processing Systems 30, pp. 5998–6008. Curran Associates, Inc. (2017). http://papers.nips.cc/paper/7181-attention-is-all-you-need.pdf
32. Wang, W., Xu, S., Xu, B.: First step towards end-to-end parametric TTS synthesis: generating spectral parameters with neural attention. In: Interspeech 2016, pp. 2243–2247 (2016). https://doi.org/10.21437/Interspeech.2016-134
33. Wang, Y., et al.: Tacotron: towards end-to-end speech synthesis. In: Proceedings Interspeech 2017, pp. 4006–4010 (2017). https://doi.org/10.21437/Interspeech.2017-1452
34. Wu, Y., et al.: Google's neural machine translation system: bridging the gap between human and machine translation. CoRR abs/ 1609.08144 (2016)
35. Zen, H., Senior, A.W., Schuster, M.: Statistical parametric speech synthesis using deep neural networks. In: IEEE International Conference on Acoustics, Speech and Signal Processing, ICASSP 2013, Vancouver, BC, Canada, May 26–31 2013, pp. 7962–7966. IEEE (2013). https://doi.org/10.1109/ICASSP.2013.6639215

Facial Expression Recognition Using Directional Gradient Local Ternary Patterns

Nahla Nour[1] and Serestina Viriri[2(✉)]

[1] College of Computer Science and Information Technology,
Sudan University of Science and Technology, Khartoum, Sudan
[2] School of Mathematics, Statistics and Computer Science,
University of KwaZulu-Natal, Durban, South Africa
viriris@ukzn.ac.za

Abstract. Extraction of human emotions from facial expression has attracted significant attention in computer vision community. There are several appearance based techniques like local binary patterns (LBP), local directional patterns (LDP), local ternary patterns (LTP) and gradient local ternary patterns (GLTP). Recently, many investigations have been done to improve these feature extraction techniques. Although GLTP has achieved an improvement in robustness to noise and illumination, it encodes image gradient in four directions and two orientations only. This paper proposes to improve GLTP to directional gradient local ternary patterns (DGLTP) by encoding image gradient on eight directions and four orientations. The eight directional Kirsch mask is used to encode the image gradient followed by dimensionality reduction using linear discriminant analysis (LDA) and AVG, MAX and MIN pooling techniques are compared for fusing facial expression features. The proposed technique was experimented on JAFFE facial expression dataset with support vector machine (SVM). The experimental results show that proposed technique improved accuracy of GLTP.

Keywords: Local binary pattern · Local directional pattern · Local ternary pattern · Gradient local ternary pattern · Directional gradient local ternary pattern

1 Introduction

In the recent years recognition of human facial expression has been an active research area in computer vision. There have been several advances in the past few years in terms of face detection and tracking, feature extraction mechanisms and the techniques used for expression classification [1], so with the increasing immersion of computers in our everyday life, the gap between computer and humans becomes increasingly apparent. Computer scientists attempt to extract social signals from human behavior to bridge the gap that separates computers from humans [2].

© Springer Nature Switzerland AG 2019
R. Chamchong and K. W. Wong (Eds.): MIWAI 2019, LNAI 11909, pp. 87–96, 2019.
https://doi.org/10.1007/978-3-030-33709-4_8

Face expression using facial images has a wide scope of applications, such as human computer interaction (HCI), emotion analysis, automated tutoring systems, smart environments, operator fatigue detection in industries, interactive video, indexing and retrieval of image and video databases, image understanding, and synthetic face animation [3]. Yet, it is a challenging task to design and develop a robust computer system for classify and recognize the face expression of the human with high accuracy. Expression plays a significant role in human communication. Facial Expression recognition is a process performed by computers and consists of three stages: facial detection and tracking, features extraction, and features classification.

Face detection has been studied extensively over the recent past years [4]. Face detection is a crucial first stage towards facial recognition technology, which is used for identifying the face from the background and extracting the data [3]. There are numerous issues faced by this technique, including different facial expressions, poses, structural components, occlusion, and various challenges relating to the image itself. There have been many face detection methods designed to identify faces within an arbitrary scene [5–9]. Nevertheless, most of these systems are only capable of detecting frontal or near-frontal views of the face. A component-based application for identifying faces from the frontal or near-frontal view was developed by Heisele et al. [10], while Rowley et al. [11] system was based on the neural network and was able to detect.

Facial feature extraction can be put into categories. For instance, can be group them into two main categories appearance-based method and geometric method. In geometrical features (local feature) is the traditional approach to modelling face processing. Features such as distances, angles, and relationships between facial parts are usually extracted [12].

The last step, classification techniques are used to recognize facial classes. There are common classifier methods such as the Artificial Neural Network (ANN) [13], Support Vector Machine (SVM) [14–16]. Decision Tree (DT) [17] and K-Nearest Neighbour (KNN) [18].

2   Literature Review and Related Work

The challenges that arise in designing a portable computer vision system include illumination variation, computation efficient and random noise. To end this, the researches can extract either local (geometric-based) features or global (appearance-based) features. In geometric based method the researches extract feature information using shape, distance, position of facial components and appearance-based feature extraction method using appearance information such as pixel intensity of face image [19]. The most appearance-based features extractions methods that are used include local binary pattern (LBP) [20–23], Local directional pattern(LDP) [20,23], local ternary pattern(LTP) [24] and Gabor wavelet transform (GWT) [20,25].

LBP was proposed by [26], which divided the facial image into local regions, it used grey-level intensity value to encode the texture of the image that leads to efficient computational. This is achieved using CK datasets 90.1 and 83 for testing 6-classes and 7-classes respectively. Nevertheless, it poorly performed in illumination variation and random noise in addition to high error rate when the background changes [27]. To solve this limitation, LDP feature is obtained by computing the edge response values in all eight directions at each pixel position and generating a code from the relative strength magnitude instead of grey-level intensity as in LBP. The result from [28] shows that 93.7 and 88.4 recognition rate where achieved for 6-classes and 7-classes respectively. Although LDP has been shown to outperform LBP but in uniform and near uniform region, it produces inconsistence pattern. LTP has been shown to overcome these limitations found in LDP by adding one more discrimination coding process. More so, the LTP uses ternary code instead of binary code in LBP and the recognition rate was observed to have improved when using 7-classes (88.9).

In recent work, the gradient local ternary pattern (GLTP) combines the advantages of the previous techniques [29]. For encoding the texture of an image GLTP uses a three-level discrimination ternary coding schema of gradient magnitude value [30] and achieved 99.3 and 97.6 recognition rate for 6-classes and 7-classes, respectively. However, most of these researches achieved very good result but still there are some drawbacks for using static global threshold that affects variation in background contrast and illumination, which reduce the accuracy rate. Furthermore, using four directions to determine the facial expressions limits the accuracy of the gradient local ternary pattern (IGLTP).

3   Methods and Techniques

In this section, details of the proposed methodology and potential improvements are given. A method for classification is discussed at the end.

3.1   Gradient Local Ternary Patterns

Ahmed and Hossain [28] proposed the gradient local ternary pattern (GLTP) which is known as a local appearance-based facial texture feature, and is used to encode the local texture of a facial expression by calculating the gradient magnitudes of local neighbourhoods within the image and quantisizing the values into three different discrimination levels. The resulting local patterns are used as facial feature descriptors. GLTP aims to address the limitations of common appearance-based features LBP [20–22] and LDP [24] by combining the advantages of the Sobel-LBP [29] and LTP [24] operators. GLTP uses more robust gradient magnitude values as opposed to grey levels with a three-level encoding scheme to discriminate between smooth and highly textured facial regions. This ensures the generation of consistent texture patterns even under the presence of illumination variation and random noise. Improved GLTP (IGLTP) enhanced the efficiency and accuracy of classification by reduce the number of features extracted from images using of a more accurate Scharr gradient operator and dimensionality reduction to reduce the size of the feature vector. Also Scharr weight matrix is anisotropic, where diagonal elements have less weights comparing with Sobel diagonal weights used in IGLTP [28].

Scharr Operator. Scharr proposed a new operator that uses an optimized filter for convolution based on minimizing the weighted mean-squared angular error in the Fourier domain [31]. The Scharr operator is as fast as the Sobel operator but provides much greater accuracy when calculating the derivatives of an image. This should result in a much more accurate representation of the gradient magnitude image. The Scharr masks for a 3×3 is defined as (Fig. 1):

$$
\begin{bmatrix} -3 & 0 & +3 \\ -10 & 0 & +10 \\ -3 & 0 & +3 \end{bmatrix}
\qquad
\begin{bmatrix} -3 & -10 & -3 \\ 0 & 0 & 0 \\ +3 & +10 & +3 \end{bmatrix}
$$

(a)   (b)

Fig. 1. (a) Scharr horizontal (b) Vertical masks (3×3)

3.2 Directional Gradient Local Ternary Patterns

A common limitation of most appearance-based methods of facial expression recognition, including GLTP, is that localization is not accurate and gives weak response to genuine edges. Unfortunately, most of these features are likely to constitute redundant information as not every region of the image is guaranteed to contain the same amount of discriminative data. GLTP also makes use of the inaccurate Scharr operator for computing the gradient magnitude image when more accurate gradient operators could have been used. In this section, improvements to the GLTP method are proposed. These include the use of a more accurate eight directional Kirsch masks is used to encode the image gradient followed by dimensionality reduction using linear discriminant analysis (LDA).

Kirsch Masks. This is a first-order derivative edge detector that get image gradient by convolving 3×3 image regions with a set of masks. Kirsch defined a nonlinear edge detector technique as:

$$
P(x, y) = \max \left\{ 1, \max_{k=0}^{7} \left[|5S_k - 3T_k| \right] \right\} \tag{1}
$$

where

$$
S_k = P_k + P_{k+1} + P_{k+2}
$$

and

$$
T_k = P_{k+3} + P_{k+4} + P_{k+5} + P_{k+6} + P_{k+7}
$$

where a in k_a is evaluated as $a = a \mod 8$ and P_k $[k = 0, 1, 2 \ldots, 7]$ are eight neighboring pixels of $P(x, y)$. Image gradient in any direction is found by convolving 3×3 image region with the respective mask M_k (Fig. 2).

Algorithm 1. Improve Gradient Local Ternary Patterns

Require: Source image i.e. pre-processing cropped face
Ensure: Vector of GLTP histograms

1. Compute horizontal G_x and vertical G_y derivative approximation of image using Scharr operator..
2. compute the gradient magnitude for each pixel of the image

$$G = \sqrt{G_x^2 + G_y^2}$$

3. Apply thredshold $\pm$ t of around center gradient value G_c in 3×3 neighborhood to determine S_{GLTP} code for the image.
4. Compute positive P_{GLTP} and negative N_{GLTP} code image representation from S_{GLTP} values.
5. Split code into $m \times n$ regions.
6. Compute positive H_{PGLTP} and negative H_{NGLTP} GLTP histogram for each region.
7. Concatenate positive H_{PGLTP} and negative H_{NGLTP} GLTP histogram for each region to form feature vector.

Algorithm 2. Directional Gradient Local Ternary Patterns

Require: Source image i.e. pre-processing cropped face
Ensure: Vector of GLTP histograms

1. Compute horizontal G_x and vertical G_y derivative approximation of image using kirsch opreator.
2. compute the gradient magnitude for each pixel of the image

$$G = \sqrt{G_x^2 + G_y^2}$$

3. compute the eight direction {N,NE,E,SE,S,SW,W,NW}
4. select
   max(E,W).max(N,S).max(NE,SW).max(NW,SE).
5. Compute positive P_GLTP and negative N_GLTP code image representation from S_GLTP values.
6. Split code into $m \times n$ regions.
7. Compute positive H_PGLTP and negative H_NGLTP GLTP histogram for each region.
8. Concatenate positive H_PGLTP and negative H_NGLTP GLTP histogram for each region to form feature vector.

$$\begin{bmatrix} -3 & -3 & 5 \\ -3 & 0 & 5 \\ -3 & -3 & 5 \end{bmatrix} \quad \begin{bmatrix} -3 & 5 & 5 \\ -3 & 0 & 5 \\ -3 & -3 & -3 \end{bmatrix} \quad \begin{bmatrix} 5 & 5 & 5 \\ -3 & 0 & -3 \\ -3 & -3 & -3 \end{bmatrix} \quad \begin{bmatrix} 5 & 5 & -3 \\ 5 & 0 & -3 \\ -3 & -3 & -3 \end{bmatrix}$$

East M_0 North East M_1 North M_2 North West M_3

$$\begin{bmatrix} 5 & -3 & -3 \\ 5 & 0 & -3 \\ 5 & -3 & -3 \end{bmatrix} \quad \begin{bmatrix} -3 & -3 & -3 \\ 5 & 0 & -3 \\ 5 & 5 & -3 \end{bmatrix} \quad \begin{bmatrix} -3 & -3 & -3 \\ -3 & 0 & -3 \\ 5 & 5 & 5 \end{bmatrix} \quad \begin{bmatrix} -3 & -3 & -3 \\ -3 & 0 & 5 \\ -3 & 5 & 5 \end{bmatrix}$$

West M_4 South West M_5 South M_6 South East M_7

Fig. 2. Kirsch edge response masks in eight directions

4   Experiment Setup

4.1   Dataset

JEFFE face expression dataset is used for experimentation, it contains 213 images of 7 facial expressions [32] (6 basic facial expressions + 1 neutral) posed by 10 Japanese female models. Each image has been rated on 6 emotion adjectives by 60 Japanese subjects. The database was planned and assembled by Michael Lyons, Miyuki Kamachi, and Jiro Gyoba. The photos were taken at the Psychology Department in Kyushu University.

4.2   Pre-processing

The image is first convert to gray-scale, face is detected using Adabost face detector proposed in [33], in some situation the face was detected used Haar-cascade face detection classifier [34, 35] and resize all the images.

4.3   Principal Component Analysis

Feature enhancement involve improving discrimination power of extracted features. This enhancement had been done using PCA for dimensionality reduction which treated as a feature enhancement process because it is aimed at reducing the number of features while still maintaining or even improving its discriminating capabilities.

4.4   Results on the JEFFE Dataset

The confusion matrix for cross-validation on the JAFFE dataset is provided in Table 1. Furthermore, the comparison has been made between proposed method and other methods in the literature. The results are summarized in Tables 2 and 3. The proposed method, Directional GLTP, shows a significant improvement in recognition accuracy over traditional GLTP on the JEFFE dataset. The largest improvement in recognition accuracy was seen during 7-class cross-validation testing.

Table 1. Confusion matrix (%) for 6 facial expression classes using DGLTP

	Joy	Surprise	Anger	Fear	Disgust	Sad	Neutral
Joy	98.4	0	0	0.5	0	0	1.1
Surprise	0	97.2	0	1.5	0.1	0	1.2
Anger	0	0.3	96.4	1.2	2.1	0	0
Fear	0	2.3	0.1	92.9	0	1.4	4.2
Disgust	0	0	0.5	0.2	98.8	0.5	1.1
Sad	0	0	0.1	4.3	4	90.5	1.1
Neutral	0	0.3	0.4	0	0	0.2	99.1

4.5 Testing Procedures

In 10-fold cross-validation the data is randomly split into k equally (or nearly equally) sized segments or folds. Then 10 iterations of training and validation are performed such that within each iteration a different fold of the data is held-out for validation while the remaining 10-1folds are used for learning. For six and seven classes of face expressions, a 10-fold cross-validation has been performed. The results show that when we use our methodology we achieved higher accuracy comparing with other methods.

Tables 2 and 3 illustrate the recognition rate of GLTP, IGLTP and our proposed DGLTP method, for six and seven class expressions in JEFFE dataset. The porposed method outperform better than traditional GLTB and IGLTP. The proposed enhancements in IGLTP such as the use of the accurate Kirsch mask operator, PCA for dimensional reduction and four more max value direction, further increase the recognition rate of IGLTP (Tables 4 and 5).

Table 2. Recognition rates (%) for 6-classes

Technique	6-classes
GLTP	74.4
IGLTP	81.7
Proposed DGLTP	**98.8%**

Table 3. Recognition rates (%) for 7-classes

Technique	7-classes
GLTP	98.9
IGLTP	99.3
Proposed DGLTP	**98.7%**

Table 4. Confusion matrix (%) for 6 facial expression classes using DGLTP

	Joy	Surprise	Anger	Fear	Disgust	Sad	Neutral
Joy	98.4%	0	0	0.5%	0	0	1.1%
Surprise	0	97.2%	0	1.5%	0.1%	0	1.2%
Anger	0	0.3%	96.4%	1.2%	2.1%	0	0
Fear	0	2.3%	0.1%	92.9%	0	1.4%	4.2%
Disgust	0	0	0.5%	0.2%	98.8%	0.5%	1.1%
Sad	0	0	0.1%	4.3%	4%	90.5%	1.1%
Neutral	0	0.3%	0.4%	0	0	0.2%	99.1%

Table 5. Facial expressions recognition rates using JEFFE dataset

Expression	6-classes	7-classes
LBP [35]	90.1 (%)	88.3 (%)
LDP [35]	93.7 (%)	88.4 (%)
LTP [35]	93.6 (%)	88.9 (%)
GLTP [35]	97.2 (%)	91.7 (%)
IGLTP [36]	99.3 (%)	97.6 (%)
DGLTP	98.8(%)	99.7 (%)

5 Conclusion

This paper proposes a facial expression recognition based on the Directional Gradient Local Ternary Patterns (DGLTP), to improve GLTP to by encoding image gradient on eight directions and four orientations. The eight directional Kirsch mask is used to encode the image gradient. The proposed technique achieved accuracy rates of 98.8% and 99.7% on 6 and 7 facial expressions classes, respectively. The experimental results show that proposed technique improved accuracy of the GLTP.

References

1. Fasel, B., Luettin, J.: Automatic facial expression analysis: a survey. Pattern Recogn. **36**(1), 259–275 (2003)
2. Tan, D., Nijholt, A.: Brain-computer interfaces and human-computer interaction. In: Tan, D., Nijholt, A. (eds.) Brain-Computer Interfaces, pp. 3–19. Springer, London (2010). https://doi.org/10.1007/978-1-84996-272-8_1
3. Zavaschi, T.H., Britto Jr., A.S., Oliveira, L.E., Koerich, A.L.: Fusion of feature sets and classifiers for facial expression recognition. Expert Syst. Appl. **40**(2), 646–655 (2013)
4. Taigman, Y., Yang, M., Ranzato, M.A., Wolf, L.: DeepFace: closing the gap to human-level performance in face verification. In: Proceedings of the IEEE Conference on Computer Vision and Pattern Recognition, pp. 1701–1708 (2014)

5. Hjelmås, E.: Face detection: a survey. Comput. Vis. Image Underst. **83**, 236–274 (2001)
6. Sung, K., Poggio, T.: Example-based learning for view-based human face detection. IEEE Trans. Pattern Anal. Mach. Intell. **20**(1), 39–51 (1998)
7. Viola, P., Jones, M.: Robust real-time object detection. In: International Workshop on Statistical and Computational Theories of Vision Modeling, Learning, Computing, and Sampling (2001)
8. Li, S., Gu, L.: Real-time multi-view face detection, tracking, pose estimation, alignment, and recognition. In: IEEE Conference on Computer Vision and Pattern Recognition Demo Summary (2001)
9. Pentland, A., Moghaddam, B., Starner, T.: View-based and modular eigenspaces for face recognition. In: Proceedings IEEE Conference Computer Vision and Pattern Recognition, pp. 84–91 (1994)
10. Heisele, B., Serre, T., Pontil, M., Poggio, T.: Component-based face detection. In: Proceedings IEEE Conference on Computer Vision and Pattern Recognition (CVPR) (2001)
11. Rowley, H., Baluja, S., Kanade, T.: Neural network-based face detection. IEEE Trans. Pattern Anal. Mach. Intell. **20**(1), 23–38 (1998). https://www. analyticsvidhya.com/blog/2017/04/comparison-betweendeep-learning-machine-learning
12. Yang, M.H., Kriegman, D.J., Ahuja, N.: Detecting faces in images: a survey. IEEE Trans. Pattern Anal. Mach. Intell. **24**(1), 34–58 (2002)
13. Shan, C., Gong, S., McOwan, P.W.: Facial expression recognition based on local binary patterns: a comprehensive study. Image Vis. Comput. **27**(6), 803–816 (2009)
14. You, H., Rumbe, G.: Comparative study of classification techniques on breast cancer FNA biopsy data. Int. J. Artif. Intell. Interact. Multimedia (2010)
15. Barrena, J.T., Valls, D.P.: Tumor Mass Detection through Gabor Filters and Supervised Pixel-Based Classification in Breast Cancer. University Rovira, Virgil (2014)
16. Refaeilzadeh, P., Tang, L., Liu, H.: Cross-validation. In: Liu, L., Özsu, M.T. (eds.) Encyclopedia of Database Systems, pp. 532–538. Springer, Boston (2009). https:// doi.org/10.1007/978-0-387-39940-9
17. Mitchell, T.M.: Machine learning. In: WCB. McGraw-Hill, Boston (1997)
18. Vaidehi, S., Vasuhi, V., et al.: Person authentication using face detection. In: Proceedings of the World Congress on Engineering and Computer Science, pp. 222–224 (2008)
19. Reis, H.T., Andrew Collins, W., Berscheid, E.: The relationship context of human behavior and development. Psychol. Bull. **126**(6), 844 (2000)
20. Suja, P., Thomas, S.M., Tripathi, S., Madan, V.K.: Emotion recognition from images under varying illumination conditions. In: Balas, V.E., Jain, L.C., Kovačević, B. (eds.) Soft Computing Applications. AISC, vol. 357, pp. 913–921. Springer, Cham (2016). https://doi.org/10.1007/978-3-319-18416-6_72
21. Sagonas, C., Tzimiropoulos, G., Zafeiriou, S., Pantic, M.: 300 Faces in-the-Wild Challenge: The first facial landmark localization Challenge. In: 2013 IEEE International Conference on Computer Vision Workshops (ICCVW), IEEE, Sydney (2013)
22. Huang, D., Shan, C., Ardabilian, M., Wang, Y., Chen, L.: Local binary patterns and its application to facial image analysis: a survey. IEEE Trans. Syst. Man Cybernet. Part C Appl. Rev. **41**(6), 765–781 (2011)
23. Jabid, T., Kabir, M.H., Chae, O.: Robust facial expression recognition based on local directional pattern. ETRI J. **32**(5), 784–794 (2010)

24. Tan, X., Triggs, B.: Enhanced local texture feature sets for face recognition under difficult lighting conditions. IEEE Trans. Image Process. **19**(6), 1635–1650 (2010)
25. Bashyal, S., Venayagamoorthy, G.K.: Recognition of facial expressions using Gabor wavelets and learning vector quantization. Eng. Appl. Artif. Intell. **21**(7), 1056–1064 (2008)
26. Singh, S.K., Chauhan, D.S., Vatsa, M., Singh, R.: A robust skin color based face detection algorithm. Tamkang J. Sci. Eng. **6**(4), 227–234 (2003)
27. Punitha, A., Kalaiselvigeetha, M.: Texture based emotion recognition from facial expression using support vector machine. Int. J. Comput. Appl. (0975–8887) **80**(5) (2013)
28. Ahmed, F., Hossain, E.: Automated facial expression recognition using gradient-based ternary texture patterns. Chin. J. Eng. **2013**, 1–8 (2013)
29. Yow, K.C., Cipolla, R.: Feature based human face detection. Image Vis. Comput. **15**(9), 713–735 (1997)
30. Hsu, C.-W., Lin, C.-J.: A comparison of methods for multiclass support vector machines. IEEE Trans. Neural Networks **13**(2), 415–425 (2002)
31. Scharr, H.: Optimal operators in digital image processing. Ph.D. thesis (2000)
32. Dhall, A., Goecke, R., Lucey, S., Gedeon, T.: Static facial expression analysis in tough conditions: data, evaluation protocol and benchmark. In: 2011 IEEE International Conference on Computer Vision Workshops (ICCV Workshops), pp. 2106–2112. IEEE, November 2011
33. Shrivastava, K., Manda, S., Chavan, P.S., Patil, T.B., Sawant-Patil, S.T.: Conceptual model for proficient automated attendance system based on face recognition and gender classification using haar-cascade, LBPH algorithm along with LDA model. Int. J. Appl. Eng. Res. **13**(10), 8075–8080 (2018)
34. Padilla, R., Costa Filho, C.F.F., Costa, M.G.F.: Evaluation of Haar cascade classifiers designed for face detection. World Acad. Sci. Eng. Technol. **64**, 362–365 (2012)
35. Ahmed, F., Kabir, M.H.: Directional ternary pattern (DTP) for facial expression recognition. In: 2012 IEEE International Conference on Consumer Electronics (ICCE). IEEE, Las Vegas (2012)
36. Holder, R.P., Tapamo, J.R.: Improved gradient local ternary patterns for facial expression recognition. EURASIP J. Image Video Process. **2017**(1), 42 (2017)

The Entity Recognition of Thai Poem Compose by Sunthorn Phu by Using the Bidirectional Long Short Term Memory Technique

Orathai Khongtum[✉], Nuttachot Promrit[✉],
and Sajjaporn Waijanya[✉]

Center of Excellence in AI and NLP, Department of Computing,
Faculty of Science, Silpakorn University, Sanam Chandra Place Campus,
Muang District, Nakhon Pathom, Thailand
{khongtum_o, promrit_n, waijanya_s}@silpakorn.edu

Abstract. The challenge of the Named Entity Recognition on the domain Thai Poem Klon-Suphap comprise of incomplete sentences, prosody, word transformation and art in language. In this article, we propose the Name Entity Recognition on the domain Thai Poem Klon-Suphap by using The Bidirectional Long Short Term network (BiLSTM) 2 models (1) BiLSTM with words embedding and (2) BiLSTM with words embedding and part of speech embedding. There were 6,216 sentences (waks) of Thai poem Phra-Aphai-Mani. The training data 4,972 sentences and testing data 1,244 sentences to recognize (1) Activity (2) Person (3) Location (4) Number (5) Body (6) Time (7) Animal and (8) Others. The experimental results of BiLSTM with words embedding and part of speech embedding showed the Precision equal 0.89, the Recall equal 0.80 and the F-measure equal 0.84. The accuracy of results is higher than BiLSTM with words embedding.

Keywords: Poem entity recognition · Bidirectional long short term memory · Thai poem · Klon-Suphap

1 Introduction

Poem is a work of art that presents the language. It has different characteristics from prose writing because it is based on grammar called prosody. he prosody is defined differently according to the type of verse. In Thailand, the most popular verse is Klon-Suphap or Klon-Pad according to it communicates with the content of a beautiful language and it has a beautiful touch position, both external and internal rhyme. According to the number of syllables in a paragraph is not too small, the poet is popular to write plays or a poem that tells a continuous story as Thai Epic. It has more suitable for playing words in order to add elegance and express the improvisation.

Phra Aphai Mani is one of Thailand's most well-known Klon-Suphap and epic poem. It was authored by the poet Sunthorn Phu; Sunthorn Phu was a honored by UNESCO [1] as a great world poet. The poem is outstanding for Sunthorn Phu employed his imagination to create an original and unique story. The story is a divided into episodes and used the characters as a narrator visually describing the whole story.

© Springer Nature Switzerland AG 2019
R. Chamchong and K. W. Wong (Eds.): MIWAI 2019, LNAI 11909, pp. 97–108, 2019.
https://doi.org/10.1007/978-3-030-33709-4_9

Furthermore, Sunthorn Phu's Klon-Suphap had more internal rhyme within verses than others. The example of the phonetic alphabets from the poem Phra Aphai Mani with internal rhymes contains both assonance such as and alliteration shown in Figs. 1 and 2.

Many mermaids swim and play in the sea.
เห็นฝูงเงือกเกลือกกลิ้งมากลางชล
Hen^5-fung^5-ngvak^3-klvak^2-kling^3-ma^1-klang^1-chon^1

Looks seem the human has tail as the fish.
คิดว่าคนมีหางเหมือนอย่างปลา
khid^4-wa^3-khon^1-mi^1-hang^5-mvan^5-jang^2-pla^1

No answer from them, so plunge and catches.
ครั้นถามไถ่ไม่พูดก็โผนจับ
Kran^4-tham^5-thaj^2-maj^3-phud^3-k@^-phon^5-cab^2

Toss and roll in the sea to catch the fish.
ดูกลอกกลับกลางน้ำปลิ้มัจฉา
du^1-kl@n^1-klab^2-klang^1-nam^4-plam^3-mad^4-cha^5

Fig. 1. The example of assonance in Klon-Suphap Poem.

Many mermaids swim and play in the sea.
เห็นฝูงเงือกเกลือกกลิ้งมากลางชล
Hen^5-fung^5-ngvak^3-klvak^2-kling^3-ma^1-klang^1-chon^1

Looks seem the human has tail as the fish.
คิดว่าคนมีหางเหมือนอย่างปลา
khid^4-wa^3-khon^1-mi^1-hang^5-mvan^5-jang^2-pla^1

No answer from them, so plunge and catches.
ครั้นถามไถ่ไม่พูดก็โผนจับ
Kran^4-tham^5-thaj^2-maj^3-phud^3-k@^-phon^5-cab^2

Toss and roll in the sea to catch the fish.
ดูกลอกกลับกลางน้ำปลิ้มัจฉา
du^1-kl@n^1-klab^2-klang^1-nam^4-plam^3-mad^4-cha^5

Fig. 2. The example of alliteration in Klon-Suphap Poem.

The assonance and alliteration will improve more dramatically attraction from the readers and make the poem has more melodiousness. Other than sensing the beauty of the poem, understanding the story lead to readers' appreciation of the poem. But when the poet composed the poem with many internal rhymes, difficult words and complex words will be appear. Reading Sunthorn Phu's poem might not too difficult for skilled poem readers but for the beginners considering that they are probably not familiar with poem terminologies and Sunthorn Phu's works have proper nouns. For example, phon^5-cab^2 (โผนจับ); jump suddenly to catch is an ancient vocabulary and non-expert readers would find it unaccustomed that they would not able to picture the story. Besides, there are some meaningless words in poems since they were intentionally modified for internal rhymes within verses and external rhymes between verses, and some of them cannot be found in dictionary. From this cause, the researcher has a NER task concept applied.

Name Entity Recognition (NER) work has been developed in multiple languages, Thai is one of the most challenging for its uniqueness; it has no spaces between words, and it has no more articles. Thai poetic language, particularly, is even more ambitious since it is in extremely complicated context and texts in verses was not completely written according to syntax. In this study, the researcher is interested in the use of Bidirectional Long Short Term network in extract name entities in Poem comparing two features between word embedding and word embedding with part of speech to create a corpus for the future Natural Language Processing work.

2   Relate Work

Nowadays, many researchers are interested in Thai Natural Language Processing. Most of the research focuses on prose text, social text, text on the web, etc. Our Center of Excellence in AI and NLP focus on poem text, We had started the project by developing the Artificial intelligence to compose Klon-suphap. We had to develop The thai poem classification [2], Thai poet identification [3] and the evaluation of thai poem's content consistency [4] to proof the machine can understanding the thai poem.

Name Entity Recognition is one of the main tasks in developing Natural language processing applied in other tasks, such as Machine translation, Information retrieval, Chat Bot and etc. NER was developed in many languages. It also has been progressed in various domain; news, academic journals, social media, medical text [5, 6], etc. and the traditionally usual way to do NER is Conditional Random Fields (CRF) [7–9]. In Thai language, use CRF [8] to identify names, locations and organization names from Best 2009 corpus comparing word-segmented data and syllable-segmented data, the study indicated syllable segmented data produced the better outcome but it could not recognize entities that were not in word lists feature, for instance abbreviation. Additionally, employed applied statistic model [10] with rules to extract people names, the finding was accurate. In Chinese, [11] introduced the comparison between neural network and CRF model to test word vectors by n-skip-gram as an input and adapted some parameters, such as word vector, window size, learning rate and etc., the finding showed the significance of word vector in neural network for increasing Recall and F1-score.

Nowadays, multiple tasks use deep learning in NER [12–14]. Deep learning's most popular architecture is Bi-directional long Short-Term Memory (BiLSTM) since it is used with sequential tasks and its ability to learn whether the text is in right or left context. NER on Twitter [12] which the language being communicated is informal, the study used word embedding, neighboring word embedding and part of speech tag as features, and the result appeared that the usage of word embedding with part of speech tag was the most effective feature. A lot of studies showed the importance of part of speech due to it being a variable giving a superior result. Moreover, Chinese NER on Telecommunication [14] domain suggested 2 module char2vec that is able to capture sematic and syntactic of alphabets and compare them with word2vec, the study concluded that char2vec + BiLSTM provided the better consequence than word2vec + BiLSTM in Chinese. This work presented BiLSTM for the poem NER, it is a new domain and nobody has done research on this matter before.

3   The Methodology

3.1   Data Preparation

This research used the klon-supharp poem title "the story of Phra Aphai Mani" total of 10 episodes (6,212 sentences) to be the data set. The Klon-supharp Structure and Prosody [3] is shown in Fig. 3. Those data is crawled from the Vajirayana digital library (https://vajirayana.org/).

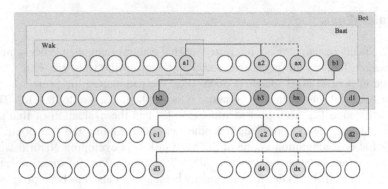

Fig. 3. Klon-supharp stracture and prosody.

Since the story of Phra Aphai Mani is a poetic drama and uses the character as a narrator to describe the whole story. To ensure the correctness of the story interpretation, we consulted the experts from the Thai Poet Contemporary Association to supervise word annotation. Therefore, the Name Entity in this research will focus on the names of persons, activities, and locations. Moreover, we are also interested in other types of Name Entity that help expand the meaning. These will describe the story for more understanding about who, what, where. The example of poem entity is shown in Fig. 4.

Many mermaids swim and play in the sea.
เห็นฝูงเงือกเกลือกกลิ้งมากลางชล
hen^5-fung^5-ngvak^3-klvak^2-kling^3-ma^1-klang^1-chon^1

Looks seem the human has tail as the fish.
คิดว่าคนมีหางเหมือนอย่างปลา
khid^4-wa^3-khon^1-mi^1-hang^5-mvan^5-jang^2-pla^1

No answer from them, so plunge and catches.
ครั้นถามได้ไม่พูดก็โผนจับ
kran^4-tham^5-thaj^2-maj^3-phud^3-k@^-phon^5-cab^2

Toss and roll in the sea to catch the fish.
ดูกลอกกลับกลางน้ำปล้ำมัจฉา
du^1-kl@n^1-klab^2-klang^1-nam^4-plam^3-mad^4-cha^5

▪ Location: words refer to place
▪ Body: words refer to anatomical parts
▪ Animal: words refer to animal
▪ Activity: words refer to states and actions of lives
▪ Person: words refer to people

Fig. 4. The example of poem entity.

To prepare the data for training sets and testing set. the poem 6,126 sentences must through word segmentation process and was annotated Parts of Speech and Name Entities including Activities, People, Places, Numbers, Bodies, Times, Animals and Other. The number of words group by Entity is shown in Fig. 5. We separated 6,212 sentences to be 4,972 sentences as training data and 1,244 sentences as testing data and store in text files. The data format in text files is shown in Fig. 6.

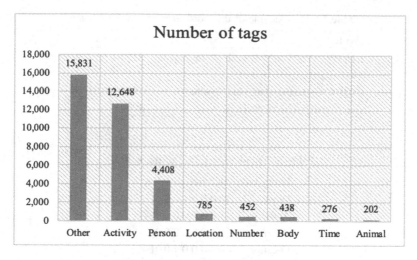

Fig. 5. The number of entity tags.

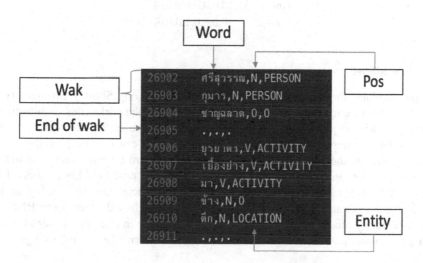

Fig. 6. The example of data format in text files.

In the Fig. 6. Each line shows the components of 1 word in the order that appears in Wak with the format [Word, POS, Entity] and shows the end of Wak with the format [.,.,.]. To prepare the input of model. First, we transform the data format from text to be the integer by mapping the word to the integer value. Next, we prepare the transformed sentence to be the same length by padding to the input and output follow the maximum length of all sentences and the maximum length of sentence in the story of Phra Aphai Mani is 10. The example of the word mapping to the integer is shown in Table 1.

In this research, we used One-hot Encoding to store the form of the answer. Thus, we transform the Entity to One-hot encoding as shown in Table 2.

Table 1. The example of the word mapping to the integer.

Words	Integer values
ศรีสุวรรณ (sri^5-su^2-wan^1, name person)	2450
กุมาร (ku^2-man^1, baby)	45
ชาญฉลาด (chan^1-cha^2-lad^2, smart)	12
มัจฉา (mad^4-cha^5, fish)	4000

Table 2. One-hot encoding of the entity.

Entity type	One-hot encoding
Other	[0,0,0,0,1,0,0,0]
Activity	[1,0,0,0,0,0,0,0]
Person	[0,1,0,0,0,0,0,0]
Location	[0,0,0,0,0,1,0,0]
Number	[0,0,0,1,0,0,0,0]
Body	[0,0,1,0,0,0,0,0]
Time	[0,0,0,0,0,0,1,0]
Animal	[0,0,0,0,1,0,0,0]

3.2   The Model

This part explains the procedure which is Bidirectional Long Short Term memory for this paper. Recurrent neural network is one of Neural Networks, this architecture is usually used with sequent information, for instance; text, voice, etc. RNN makes use of previous data's orders to calculatedly re-input data so it can capture preceding information. However, the architecture failed to learn long-term dependencies; it could not sufficiently make use of previous data. To solve this problem, Long Short Term Memory was added, it could learnt long-term dependencies and sustainably capture information for advanced usage. LSTM's structure is slightly different from RNN's, it contains memory blocks called cells which has gates to be in charge; the gates includes input gate, output gate, forget gate, and memory cell calculating as followings:

$$f_t = \sigma(w_f[h_{t-1}, x_t] + b_f) \tag{1}$$

$$i_t = \sigma(w_i[h_{t-1}, x_t] + b_i \tag{2}$$

$$c_t = f_t * c_{t-1} + i_t * \tanh(w_c[h_{t-1}, x_t] + b_c) \tag{3}$$

$$o_t = \sigma(w_o[h_{t-1}, x_t] + b_o) \tag{4}$$

$$h_t = o_t * \tanh(c_t) \tag{5}$$

where σ are the sigmoid function, x are the input vector, f, i, c, o are forget gate, input gate, memory cell and output gate, h are the output information, w and b are the weight and bias, respectively. LSTM can only make use of preceding context, but BILSTM linking with LSTM can learn from both left $\overleftarrow{h_t}$ preceding context and right $\overrightarrow{h_t}$ future context to predict the proceeding information so that it could better access context. In our approach, we experimented including 2 models with different input layer. The 1st model we used only word feature (BiLSTM with word feature) is shown in Fig. 7 and the 2nd model we used both word feature and word concatenate part of speech (BiLSTM with word concatenate part of speech) is shown in Fig. 8. The characteristic of part of speech, we used 4 tags as Noun, Pronoun, Verb and Other.

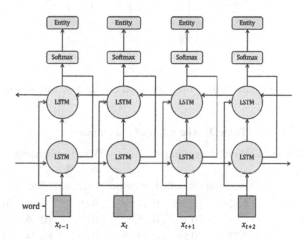

Fig. 7. The model of BiLSTM with word feature.

4 The Experiment and Result

In the experiment, the data were divided into training set 80% and testing set 20%. The training set data was 4,972 sentences in a total amount and testing set was 1,244. The features used in the experiment including word embedding and word embedding with part of speech embedding. Our models are trained using Bidirectional Long Short term memory. We used different parameters. The parameters included the number of dimension in the embedding layer, dropout value and Batch Normalization layer. The measurement used Precision, Recall and F-measure.

The experiment results in Table 3 shows the comparison of the Number of dimension. The model structure consists of 4 layers including input layer, embedding layer, BiLSTM layer and output layer. The input layer is the word embedding feature. The number of word dimensions used in this layer is 100, 150 and 200 dimensions. Besides, we fed the features input layer which combined between word embedding and part of speech embedding in the input layer with the part of speech dimension is 20.

Based on the results of the experiment which has shown in Table 3, the performance of the model is not satisfactory. Therefore, we adjusted the parameter to improve the results by using dropout value = 0.2 and Batch Normalization layer with both features. In this model, we used batch size = 512 and used optimizer function is Adam. We set

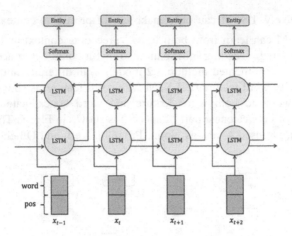

Fig. 8. The model of BiLSTM with word concatenate part of speech.

Table 3. The result of experiments.

Number of dimension	Word embedding			Word embedding + pos embedding		
	Precision	Recall	F-measure	Precision	Recall	F-measure
100	0.72	0.59	0.63	0.86	0.70	0.74
150	0.83	0.70	0.75	0.86	0.66	0.71
200	0.77	0.58	0.63	0.85	0.74	0.78

the learning rate to 0.0001 and used cross-entropy loss function. The dimension of word embedding is 100, 150 and 250 and the part of speech embedding is 20.

Table 4 shows the comparative performance between word embedding feature and word embedding with part of speech embedding feature. The best value of both features presented value of precision, recall and F-measure. For embedding feature, the best value is 0.89, 0.76 and 0.81 respectively and 0.89, 0.80 and 0.84 of word embedding with part of speech embedding feature. The results in the table indicates the use of word embedding with part of speech helped improving F-measure value 3%, the value of recall increases 4%.

Table 4. The result of two features with dropout = 0.2.

Number of dimension	Word embedding			Word embedding + pos embedding		
	Precision	Recall	F-measure	Precision	Recall	F-measure
100	0.87	0.76	0.81	**0.89**	**0.80** (+4)	**0.84** (+3)
150	0.88	0.75	0.80	0.90	0.78	0.83
200	**0.89**	**0.76**	**0.81**	0.81	0.77	0.78

We have shown the loss from the training model in Fig. 9. Loss of model. and the precision, recall, and F-measure of each poem entity in Table 5. Precision, Recall and F-measure of each poem entity. Observation the increased score of activity entity is clearly seen when adding part of speech feature. The precision increased by 7%, recall increased by 9% and F-measure increased by 8%.

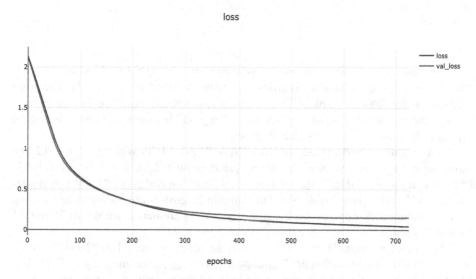

Fig. 9. Loss of model.

Table 5. Precision, Recall and F-measure of each poem entity.

Entity	Word embedding			Word + pos embedding		
	Precision	Recall	F-measure	Precision	Recall	F-measure
Time	0.91	0.60	0.72	0.91	0.62	0.74
Person	0.93	0.89	0.91	0.91	0.90	0.91
Number	0.91	0.88	0.90	0.96	0.90	0.93
Location	0.89	0.64	0.75	0.84	0.67	0.75
Body	0.75	0.70	0.72	0.81	0.77	0.79
Animal	1.00	0.61	0.76	0.88	0.64	0.74
Activity	**0.86**	**0.86**	**0.86**	**0.93** (+7)	**0.95** (+9)	**0.94** (+8)
Other	0.84	0.88	0.86	0.91	0.92	0.91

From the experimental results, we analyzed the results in the test data set. In our work, this tagging will help unskilled readers for better interpretation the poem. We show an example of sentence that predict correctly in the Table 6.

Table 6. Example of result.

Word	ศรีสุวรรณ Srisuwan	นั้น -	มา Come to	ค้าง Stay	อยู่ Live	ปรางค์มาศ Pagoda
Word embedding	Person √	Other √	Activity √	Activity √	Activity √	Location √
Word embedding with part of speech embedding	Person √	Other √	Activity √	Activity √	Activity √	Location √

From the Table 6. The sentence is si^5-su^2-wan^1-nan^4-ma^1-khang^4-ju^2-prang^1;mad^3 (ศรีสุวรรณ-นั้น-มา-ค้าง-อยู่-ปรางค์มาศ); Srisuwan stayed at the pagoda. The word "ปรางค์มาศ" is the difficult word in Thai language and not appear in the Thai dictionary because the root word is "ปรางค์" that meant to "pagoda" but the poet transformed to be "ปรางค์มาศ" because of the poem melodiousness.

To represent the performance of BiLSTM + Word + POS with dropout = 0.2, we show the example of sentence that wrong predicts by BiLSTM + Word in Table 7. From the result in Table 7, the sentence is klxng^3-naxn^5-ng#j^1-du^1-duang^1-bub^2-pha^5 (แกล้ง-แหงนเงย-ดู-ดวง-บุปผา) the meaning in English is pretending to be look-ing up at a bouquet of flowers. In this example, the wrong predicts is word naxn^5-ng#j^1 (แหงนเงย: Turn up).

The Another example in Table 8, the sentence is tam^1-ka^2-sad^2-svb^2-wong^1-dam^1-rong^1-sa^2-than^5 (ตาม-กษัตริย์-สืบวงศ์-ดำรงสถาน) the meaning in English is The typical of the king in the succession and rule of the country. the wrong predicts are words svb^2-wong^1 (สืบวงศ์: Succeed) and dam^1-song^1-sa^2-than^5 (ดำรงสถาน: reign).

From the result in Tables 7 and 8, The model BiLSTM + Word + POS with dropout = 0.2 can predict the type of entity correctly when adding part of speech. We can describe about the wrong predicts words by the BiLSTM + Word model. The reason is they are combined words and difficult words because of they are disappear in the Thai dictionary. The poem by professional poets are often used difficult words for increasing more beautiful language and melodic and they will compose the poem by strict prosodic and ignore the grammar of the language. Whenever we added a part of speed embedding to the model, the model performance has been increased and it can be predicted the Entity of difficult words.

Table 7. Example of result.

Word	แกล้ง Pretend	แหงนเงย Turn up	ดู See	ดวง -	พวง Bunch	บุปผา Flower
Word embedding	Activity √	Other ×	Activity √	Other √	Other √	Other √
Word embedding with part of speech embedding	Activity √	Activity √	Activity √	Other √	Other √	Other √

Table 8. Example of result.

Word	ตาม Pretend	กษัตริย์ King	สืบวงศ์ Succeed	ดำรงสถาน Reign
Word embedding	Other √	Person √	Other ×	Activity ×
Word embedding with part of speech embedding	Other √	Person √	Activity √	Activity √

Notice the model which using "word embedding" the system ability for remembering is quite pretty good but increasing the part of speech to have better system ability which demonstrates the increasing score. For example, the sentence "แลดูปีที่เป่าเล่าก็หาย" and to consider the word "เล่า" it can be in many role, focusing in this context it is not verb that meant to activity but in word embedding it has been predicted as a activity but in Word + Pos it was "Other".

5   Conclusion

This paper uses Bidirectional long Short Term Memory (BiLSTM) for poem entity recognition. There are 7 types of entities including Activity, Person, Location, Number, Body, Time, Animal, and Others. We compared two features between word embedding and word embedding with part of speech. The word embedding with part of speech feature shows the best performance of precision, recall, and F-measure results 0.89, 0.80 and 0.84 respectively. However, we would like to propose other features for increasing the model efficiency in the future. The Thai poem Name Entity recognition is challenge work. The future propose is not only improving the performance of the model but we will apply to other related work such as the automate interpret Thai poem to preserve and disseminate the beauty of Thai language arts in the age of technology.

References

1. Phu, S.: ThingsAsian: Thailand's Shakespeare? http://thingsasian.com/story/thailands-shakespeare-sunthorn-phu
2. Promrit, N., Waijanya, S.: Convolutional neural networks for thai poem classification. In: Cong, F., Leung, A., Wei, Q. (eds.) ISNN 2017. LNCS, vol. 10261, pp. 449–456. Springer, Cham (2017). https://doi.org/10.1007/978-3-319-59072-1_53
3. Waijanya, S., Promrit, N.: The poet identification using convolutional neural networks. In: Meesad, P., Sodsee, S., Unger, H. (eds.) IC2IT 2017. AISC, vol. 566, pp. 179–187. Springer, Cham (2018). https://doi.org/10.1007/978-3-319-60663-7_17
4. Promrit, N., Waijanya, S., Thaweesith, K.: The evaluation of Thai poem's content consistency using siamese network. In: Proceedings of the 2019 3rd International Conference on Natural Language Processing and Information Retrieval, pp. 115–120. ACM, New York (2019). https://doi.org/10.1145/3342827.3342855

5. Tong, F., Luo, Z., Zhao, D.: Using deep neural network to recognize mutation entities in biomedical literature. In: 2018 IEEE International Conference on Bioinformatics and Biomedicine (BIBM), pp. 2329–2332 (2018)
6. Zeng, D., Sun, C., Lin, L., Liu, B.: LSTM-CRF for drug-named entity recognition. Entropy **19**, 283 (2017)
7. Taufik, N., Wicaksono, A.F., Adriani, M.: Named entity recognition on Indonesian microblog messages. In: 2016 International Conference on Asian Language Processing (IALP), pp. 358–361 (2016). https://doi.org/10.1109/IALP.2016.7876005
8. Tirasaroj, N., Aroonmanakun, W.: Thai named entity recognition based on conditional random fields. In: 2009 Eighth International Symposium on Natural Language Processing, pp. 216–220 (2009). https://doi.org/10.1109/SNLP.2009.5340913
9. Yang, X., Huang, H., Xin, X., Liu, Q., Wei, X.: Domain-specific product named entity recognition from Chinese microblog. In: 2014 Tenth International Conference on Computational Intelligence and Security, pp. 218–222 (2014). https://doi.org/10.1109/CIS.2014.73
10. Saetiew, N., Achalakul, T., Prom-on, S.: Thai person name recognition (PNR) using likelihood probability of tokenized words. In: 2017 International Electrical Engineering Congress (iEECON), pp. 1–4 (2017). https://doi.org/10.1109/IEECON.2017.8075816
11. Wang, G., Cai, Y., Ge, F.: Using hybrid neural network to address Chinese named entity recognition. In: 2014 IEEE 3rd International Conference on Cloud Computing and Intelligence Systems, pp. 433–438 (2014)
12. Rachman, V., Savitri, S., Augustianti, F., Mahendra, R.: Named entity recognition on Indonesian Twitter posts using long short-term memory networks. In: 2017 International Conference on Advanced Computer Science and Information Systems (ICACSIS), pp. 228–232 (2017)
13. Suriyachay, K., Sornlertlamvanich, V.: Named entity recognition modeling for the Thai language from a disjointedly labeled corpus. In: 2018 5th International Conference on Advanced Informatics: Concept Theory and Applications (ICAICTA), pp. 30–35 (2018)
14. Wang, Y., Xia, B., Liu, Z., Li, Y.J., Li, T.: Named entity recognition for Chinese telecommunications field based on Char2Vec and Bi-LSTMs. In: 2017 12th International Conference on Intelligent Systems and Knowledge Engineering (ISKE), pp. 1–7 (2017)

Generation of Efficient Rules
for Associative Classification

Chartwut Thanajiranthorn$^{(\boxtimes)}$ and Panida Songram

Department of Computer Science, Faculty of Informatics Mahasarakham University,
Mahasarakham, Thailand
{60011260501,panida.s}@msu.ac.th

Abstract. Associative classification is a mining technique that integrates classification and association rule mining for classifying unseen
data. Associative classification has been proved that it gives more accurate than traditional classifiers and generates useful rules which are easy
to understand by a human. Due to inheriting from association rule mining, associative classification has to face a sensitive of minimum support
threshold that a huge number of rules are generated when a low minimum
support threshold is given. Some of the rules are not used for classification and need to be pruned. To eliminate unnecessary rules, this paper
proposes a new algorithm to find efficient rules for classification. The
proposed algorithm directly generates efficient rules. A vertical data representation technique is adopted to avoid the generation of unnecessary
rules. Our experiments are conducted to compare the proposed algorithm with well-known algorithms, CBA and FACA. The experimental
results show that the proposed algorithm is more accurate than CBA
and FACA.

Keywords: Associative classification · Class Association Rule · A
vertical data representation · Classification

1 Introduction

Associative classification (AC) is a supervised learning classification technique
which was proposed by Lui et al. [11]. It integrates two important data mining
techniques together, association rules mining and classification. Association rules
mining aims to find a relationship between data. Classification aims to predict
unseen data and assigns a class label. AC focus to find a Class Association Rules
(CARs) in form *itemset* → *c* where itemset is a set of attribute values, *c* is a
class label. AC has been improved that it produces more accurate than other
traditional classifiers [19]. In addition, a CAR is an If-Then rule which is easy
to understand by users. Therefore, AC is adapted in many fields i.e. phishing
website detection [1,2,7], heart disease prediction [9,16], groundwater detection
[8], and detection of low quality information in social networks [20].

© Springer Nature Switzerland AG 2019
R. Chamchong and K. W. Wong (Eds.): MIWAI 2019, LNAI 11909, pp. 109–120, 2019.
https://doi.org/10.1007/978-3-030-33709-4_10

Li et al. [10] presented the Classification Based on Multiple Association Rules algorithm (CMAR). The CMAR algorithm generates CARs based on the FP-Growth algorithm. It adopted the Frequent Pattern Tree structure (FP-Tree) to store frequent items and classes. First, it scans a database to find the frequent items and then sorts them in F-list. Each transaction is scanned to create a FP-Tree. The CMAR algorithm divides subset in the FP-Tree to search frequent ruleitems and then add the frequent ruleitems to CR-Tree according to their frequencies. It classifies based on the weighted Chi-square analysis. It prunes unnecessary ruleitems based on confidence, correlation and database coverage. Based on the experimental results, they are shown that the CMAR algorithm is more efficient and scalable than other the CBA and C4.5 algorithms.

Although AC is widely used in many fields and efficient classification, it has to face the generation of a large number of CARs because of association rule mining inheritance. Moreover, some of CARs are unnecessary for classification. Some of algorithms, CBA [11], CAR-Miner [13], PCAR [17] and FCBA [5], attempt to create rules that are effective for classification, but they must also create candidate rules to determine whether the rules can be used for classification. Creating candidate rules spend a lot of computation times and a large number of rules are generated. Hadi et al. [8] have proved that more than 100 million candidate 3-ruleitems are generated from 1,000 frequent 2-ruleitems. Nguyen and Nguyen [13] showed that the number of 4 million candidate ruleitems have been generated when the minimum support threshold is set to 1%. Furthermore, Abdelhamid et al. [3] reported that the minimal process of candidate generation is still a challenging work because it has effective in regards to training time, I/O overheads and memory usage.

In this paper, we propose a new algorithm, called the Efficient Class Association Rule Generation Algorithm (ECARG). The ECARG algorithm is proposed a new method to directly generate a small number of efficient CARs for classification. A dataset is represented in a vertical data format [14] which each ruleitem contains transaction ids. Support and confidence values are easily calculated by using the vertical data format. An efficient CAR is directly found and added to a classifier. Then, transaction ids containing the CAR are not used to find the next CARs so that redundant CARs are not generated. Therefore, the ECARG algorithm early prunes ruleitems during CARs generation process.

The remaining sections of the paper are organized as follows. Section 2 presents related work of AC. Section 3 presents the basic definitions. The proposed algorithm is introduced in Sect. 4. Section 5 discusses the experimental results. The paper is concluded in Sect. 6.

2   Related Work

Various algorithms have been proposed for AC. Lui et al. introduced the CBA algorithm which integrated association rule mining and classification. The process of the CBA algorithm is divided into two steps. First, CARs are generated based on the Apriori algorithm [4]. Second, they are sorted and the redundant rules are pruned. The second step is an important process for selecting efficient CARs in a classifier. The CBA algorithm was proved that it produces a lower error rate than C4.5 [15]. Unfortunately, the CBA algorithm has to face the problem of candidate generation due to base on Apriori manner. It finds CARs from all possible candidate rules at each level so that a large number of candidate rules is generated. After associative classification was introduced, many different algorithms were proposed such as CMAR, eMCAC, FACA, PCAR, FCBA and ACPRISM.

Abdelhamid [1] proposed an Enhanced Multi-label Classifiers based Associative Classification (eMCAC) for phishing websites detection. The eMCAC algorithm enhances the MCAC algorithm [12] by applies a vertical data format to represented datasets. The eMCAC algorithm generates multi-label rules that attribute values are connected with more than one class. The support and the confidence and support values for a multi-label rule are calculated based on the average confidence and support values of all classes. The rules in classifier are sorted to ensure that the rules with high confidence and support values are given higher priority. The class is assigned to the test data if attribute values are fully matched with the rule's antecedent. The accuracy is evaluated and shown that the eMCAC algorithm outperforms CBA, PART, C4.5, jRiP and MCAR.

Hadi et al. [7] proposed the FACA algorithm to detect phishing websites. It applics a Diffset [21] to discover CARs. The FACA algorithm discovers k-ruleitems by extending frequent (k-1)-ruleitems. Then, ruleitems are ranked according to the number of itemset in ruleitem, confidence, support, and occurrence. To predict unseen data, the FACA algorithm utilizes All Exact Match Prediction Method. The method matches unseen data with all CARs in classifier. Next, unseen data is assigned to class label with the highest count. The FACA algorithm outperforms CBA, CMAR, MCAR, and ECAR [6].

Song and Lee introduced Predictability-Based Collective Class Association Rule algorithm (PCAR). The PCAR algorithm uses a cross-validation between test dataset and train dataset to calculate a predictability value of CARs. First, PCAR adapts the Eclat algorithm to discover CARs from a dataset. Next, it calculates an average support of CARs from each fold of testing dataset. Then, CARs are ranked according to predictive value, confidence, support, rule antecedent length and rule occurrences, respectively. Finally, the full-matching method is applied to assigns a class label for unseen data. The PCAR algorithm is compared with C4.5, Ripper, CBA and MCAR on the accuracy and shown that it outperforms the others.

Alwidian et al. proposed the WCBA algorithm to overcome the problem of most associative classification algorithms which determines the importance of rules based on support and confidence values. The WCBA assumes that every attributes are equally important regardless of real-world application. For example, medical work determines information affecting the prediction results. The weight of an attribute is calculated to indicate its importance. In addition, CARs are prior sorted by using the harmonic mean which is an average value between support and confidence. The WCBA algorithm is more significantly accurate than CBA, CMAR, MCAR, FACA, and ECBA. However, the WCBA algorithm generates CARs based on Apriori that scan database many times.

From the previous algorithms, they were proposed for associative classification with high ability of rules for prediction. However, most of the algorithms produce k-ruleitems from (k-1)-ruleitems. They have to calculate support when a new ruleitems is found. To calculate support, they have to scan database with all transactions. Moreover, a huge number of candidate CARs are generated and needed for pruning process to reduce unnecessary CARs. Unlike the previous algorithm, we propose a method to directly generate efficient CARs for prediction so that the pruning and sorting processes do not need in the proposed algorithm. Moreover, a simple set theories, intersection and set different are adapted to calculate easily support, confidence and remove unnecessary ruleitems.

3 Basic Definitions

Let $A = \{a_1, a_2, ..., a_m\}$ is a finite set of all attributes in dataset. $C = \{c_1, c_2, ..., c_n\}$ is a set of classes, $g(x)$ is a set of transactions containing itemset x and $|g(x)|$ is the number of transactions containing x.

Definition 1. *An item can be described as an attribute a_i containing a value v_j, denote as $\langle (a_i, v_j) \rangle$.*

Definition 2. *An itemset is the set of items, denoted as $(a_{i1}, v_{i1})\rangle$, $(a_{i2}, v_{i2}), ..., (a_{ik}, v_{ik})$.*

Definition 3. *A ruleitem is of the form $\langle itemset, c_j \rangle$, which represents an association between itemsets and class in dataset, basically it is represented in form itemset $\rightarrow c_j$.*

Definition 4. *The length of ruleitem is the number of items, denoted as $k -$ ruleitem.*

Definition 5. *Absolute support of ruleitem r is the number of transactions containing r, denoted as $sup(r)$. The support of r can be found from Eq. 1.*

$$sup(r) = |g(r)| \tag{1}$$

Definition 6. *Confidence of ruleitem $\langle itemset, c_j \rangle$ is the ratio of the number of transactions that contain itemset in c_j and the number of transaction contain the itemset, and is calculated by Eq. 2*

$$conf(\langle itemset, c_j \rangle) = \frac{|g(\langle itemset, c_j \rangle)|}{|g(itemset)|} \times 100 \tag{2}$$

Definition 7. *Frequent ruleitem is a ruleitem whose support is not less than the minimum support threshold (minsup).*

Definition 8. *Class association rule (CAR) is a frequent ruleitem whose confidence is not less than the minimum confidence threshold (minconf).*

4 The Proposed Algorithm

In this section, we propose a new algorithm for associative classification. The proposed algorithm includes 4 phases: (1) 1-ruleitem generation, (2) redundant rules removal, (3) ruleitem extension, and (4) default rule creation. Each step will be explained in the subsection with an example. To show an example, we use the weather dataset, as shown in Table 1. The minimum support and confidence thresholds are set to 3 and 90%, respectively.

Table 1. Weather dataset.

Transaction ID	Outlook	Temperature	Humidity	Windy	Play
1	Sunny	Hot	High	False	No
2	Sunny	Hot	High	True	No
3	Overcast	Hot	High	False	Yes
4	Rainy	Mild	High	False	Yes
5	Rainy	Cool	Normal	False	Yes
6	Rainy	Cool	Normal	True	No
7	Overcast	Cool	Normal	True	Yes
8	Sunny	Mild	High	False	No
9	Sunny	Cool	Normal	False	Yes
10	Rainy	Mild	Normal	False	Yes
11	Sunny	Mild	Normal	True	Yes
12	Overcast	Mild	High	True	Yes
13	Overcast	Hot	Normal	False	Yes
14	Rainy	Mild	High	True	No

4.1 Efficient 1-Ruleitem Generation

In the first phase, 1-ruleitem is generated on vertical data format which is easy to calculate support and confidence. The vertical data format [14] represents associated transaction ids of 1-ruleitems as shown in Table 2. The set of the transaction ids of 1-ruleitems easily find from $|g(itemset) \cap g(c_k)|$, which is the support. If the support of 1-ruleitem is less than the minimum support, the 1-ruleitem will be not further extended with other items. The last 2 rows of Table 2 show the support and confidence of ruleitems, respectively.

From Table 1, sunny value in outlook occurs in transaction ids 1, 2, 8, 9 and 11, denoted as $g(\langle Outlook, Sunny \rangle) = \{1, 2, 8, 9, 11\}$. Class Yes is in transaction ids 3,4,5,7,9,10,11,12, and 13, denoted as $g(Yes) = \{3, 4, 5, 7, 9, 10, 11, 12, 13\}$ while class No is in transaction ids 1, 2, 6, 8, and 14, denoted as $g(No) = \{1, 2, 6, 8, 14\}$. Transaction ids containing $\langle Outlook, Sunny \rangle \rightarrow Yes$ is $g(\langle Outlook, Sunny \rangle) \cap g(Yes) = \{1, 2, 8, 9, 11\} \cap \{3, 4, 5, 7, 9, 10, 11, 12, 13\} =$

$\{9, 11\}$, so the supports of $\langle Outlook, Sunny \rangle \rightarrow Yes$ is 2. Transaction ids containing $\langle Outlook, Sunny \rangle \rightarrow No$ is $g(\langle Outlook, Sunny \rangle) \cap g(No) = \{1, 2, 8, 9, 11\} \cap \{1, 2, 6, 8, 14\} = \{1, 2, 8\}$, so the supports of $\langle Outlook, Sunny \rangle \rightarrow No$ is 3. We can see that, the rule $\langle Outlook, Sunny \rangle \rightarrow Yes$ will not be extended because its support is less than the minimum support threshold.

The confidences of the remaining rules are found and easily calculated from $\frac{|g(itemset \rightarrow c_k)|}{|g(itemset)|} \times 100$. If the confidence of the rule is 100%, the rule will be added to the classifier. If not, it will be considered to extend in the next phase. For example, the confidence of $\langle Outlook, Sunny \rangle \rightarrow No$ can easily find from $\frac{|g(1,2,8)|}{|g(1,2,8,9,11)|} \times 100 = \frac{3}{5} \times 100 = 60\%$. The confidence of the rule $\langle Outlook, Sunny \rangle \rightarrow No$ is not 100% so it will be extended. While the confidence of $\langle Outlook, Overcast \rangle \rightarrow Yes$ is $\frac{|g(3,7,12,13)|}{|g(3,7,12,13)|} \times 100 = \frac{5}{5} \times 100 = 100\%$ so it is the first CAR adding to the classifier.

Table 2. The rules which passed minimum support threshold (white background cell).

	Outlook						Temperature						Humidity				Windy			
	Sunny		Overcast		Rainy		Hot		Mild		Cool		High		Normal		True		False	
	Y	N	Y	N	Y	N	Y	N	Y	N	Y	N	Y	N	Y	N	Y	N	Y	N
	9 11	1 2 8	3 7 12 13		4 5 10	9	3 13	1 2	4 10 11 12	8 14	5 7 9	6	3 4 12	1 2 8 14	5 7 9 10 11 13	6	7 11 12	2 6 14	3 4 5 9 10 13	1 8
sup	2	3	4	0	3	1	2	2	4	2	3	1	3	4	6	1	3	3	6	2
conf		60	100		60				67		75		43	57	86		50	50	75	

4.2   Redundant Rule Removal

After finding a ruleitem with 100% of confidence, the transaction ids containing the ruleitem will be removed to reduce the unnecessary CARs generation. For example, Table 1 we can see that if $\langle (Outlook, Overcast) \rangle$ is found, the class will be actually Yes. Therefore we do not need to consider any item occurs in the same transaction ids of $\langle (Outlook, Overcast) \rangle$. To remove unnecessary transaction ids, set difference is adopted. Let r_i is a CAR with 100% of confidence and T is a set of ruleitems in the same class of r_i. For all $r_j \in T$, the new transaction ids of r_j is $g(r_j) = g(r_j) - g(r_i)$. For example, in Table 2, the CAR $g(\langle Outlook, Overcast \rangle \rightarrow Yes) = \{3, 7, 12, 13\}$ and $g(\langle Humidity, High \rangle \rightarrow Yes) = \{3, 4, 12\}$. The new transaction ids of $g(\langle Humidity, High \rangle \rightarrow Yes) = g(\langle Humidity, High \rangle \rightarrow Yes) - g(\langle Outlook, Overcast \rangle \rightarrow Yes) = \{3, 4, 12\} - \{3, 7, 12, 13\} = \{4\}$. The new transaction ids of all ruleitems are shown in Table 3.

Table 3. The remained transaction ids after generating the first CAR.

	Outlook						Temperature						Humidity				Windy			
	Sunny		Overcast		Rainy		Hot		Mild		Cool		High		Normal		True		False	
	Y	N	Y	N	Y	N	Y	N	Y	N	Y	N	Y	N	Y	N	Y	N	Y	N
	9	1			4	9	1	4	8	5	6		4	1	5	6	11	2	4	1
	11	2			5		2	10		9				2	9			6	5	8
		8			10			11						8	10			14	9	
														14	11				10	
sup	2	3	0	0	3	1	2	2	3	2	2	1	1	4	4	1	1	3	4	2
conf	60				60				60				80		80		75		67	

4.3 Ruleitem Extension

The ruleitem r with highest confidence will be first considered to extend in breadth first search manner. It will be combined with other ruleitem in the same class until the new rule has 100% of confidence. If r_i is extended with r_j to be r_{new} and $g(r_j) \subseteq g(r_i)$ then $conf(r_{new}) = 100\%$.

For example, in Table 3, $g(\langle Humidity, High \rangle \to No)$ has the highest confidence and the $g(\langle Outlook, Sunny \rangle \to No) = \{1, 2, 8\}$ is subset of $g(\langle Humidity, High \rangle \to No) = \{1, 2, 8, 14\}$ so that the new rule $g(\langle Outlook, Sunny \rangle, \langle Humidity, High \rangle \to No)$ is found with 100% of confidence. Then $g(\langle Outlook, Sunny \rangle, \langle Humidity, High \rangle \to No)$ is stopped to extend. For 2-ruleitem extended from $\langle Outlook, Sunny \rangle \to No$, there is only one rule with 100% of confidence and it is added to the classifier as the second CAR.

After the second CAR is added to the classifier, the transaction ids associated with the CAR will be removed as explained in Sect. 4.2. The remaining transaction ids are shown in Table 4.

Table 4. Transaction ids after generating the second CAR.

	Outlook						Temperature						Humidity				Windy			
	Sunny		Overcast		Rainy		Hot		Mild		Cool		High		Normal		True		False	
	Y	N	Y	N	Y	N	Y	N	Y	N	Y	N	Y	N	Y	N	Y	N	Y	N
	9				4	9			4	14	5	6	4	14	5	6	11	6	4	
	11				5				10						9				9	
					10				11						10				9	
															11				10	
sup	2				3	1			3	1	2	1	1	1	4	1	1	2	4	
conf					75				75						80				100	

In Table 4, the rule $\langle Windy, False \rangle \to Yes$ has 100% of confidence and it is added to classifier as the third CAR and removes associated transaction ids. Finally, there no any ruleitem pass minimum support threshold as shown in Table 5, the CAR generation will be stopped.

Table 5. Transaction ids after generating the third CAR.

	Outlook						Temperature						Humidity				Windy			
	Sunny		Overcast		Rainy		Hot		Mild		Cool		High		Normal		True		False	
	Y	N	Y	N	Y	N	Y	N	Y	N	Y	N	Y	N	Y	N	Y	N	Y	N
	11								11	14		6	14	11	6		11	6 14		
sup	1								1	1		1	1	1	1		1	2		
conf																				

4.4 Default Rule Creation

The proposed algorithm continues to build a default CAR for adding to the classifier. In this step, the class that appears the most relevant transaction ids is selected as the default CAR. From Table 5, the remaining transaction ids is relevant to class 'No' the most, so that the default class is No. Finally, all CARs in the classifier is shown in Table 6.

Table 6. All CARs.

Rule ID	Rule
R1	$\langle Outlook, Overcast \rangle \rightarrow Yes$
R2	$(\langle Humidity, High \rangle, \langle Outlook, Sunny \rangle) \rightarrow No$
R3	$\langle Windy, False \rangle \rightarrow Yes$
Default	No

5 Experimental Setting and Result

All the experiments are performed on a 2.3 GHz Intel Core i3-6100u processor with 8 GB DDR4 main memory, running Microsoft Windows 10. We implemented FACA and ECARG algorithms in java and used the CBA algorithm in WEKA data mining software. We have performed an experiment to evaluate the accuracy and number of CARs. Our algorithm is compared with the CBA and FACA algorithms. In the experiments, the minimum support threshold is set to 2% and the minimum confidence threshold is set to 50%. The thresholds are a set according from [5,8,18]. Three algorithms are tested with 13 datasets from UCI Machine Learning Repository. The characteristics of the experimental datasets are shown in Table 7. Table 7 shows the number of attributes (exclude a class label), the number of class labels and the number of instances in each data set.

The experimental results are shown in Tables 8 and 9. Table 8 reports the accuracy of the CBA, FACA, and ECARG algorithms. It is clear that the ECARG algorithm performs on average well when comparing to the CBA and FACA algorithms. It gives higher accuracy than CBA and FACA by 4.48% and

Table 7. Characteristics of experiment dataset.

Data sets	#of attributes	#of classes	Instances
Breast	11	2	699
Cars	6	4	1,728
Contact-lenses	4	3	24
Diabetes	7	2	768
Iris	4	3	150
Labor	17	2	57
Lymph	18	4	148
Mushroom	22	2	8,214
Post-operative	9	4	90
Tic-tac-toe	9	2	958
Vote	16	2	435
Wined	13	3	178
Zoo	17	7	101

4.22% respectively. This gain has been resulted from the methodology to find the most efficient rule in each iteration and eliminate redundant rules simultaneously.

To be more precise, the proposed algorithm gives the highest accuracy in 9 of 13 datasets. We further analyzed the win-tie-lost records. Based on the Table 8, win-tie-lost records of the proposed algorithm against CBA and FACA in term of accuracy are 12-0-1 and 11-0-2, respectively.

Table 9 shows the average number of CARs from the CBA, FACA and ECARG algorithms using 10-fold cross-validation. The result shows that our algorithm has derived a smaller number of rules than the CBA algorithm. In particular, the ECARG algorithm generates 16 CARs on average against 13 datasets, while the CBA algorithm derives 21 rules on average and the FACA algorithm derives 14 rules on average. We can see that the ECARG algorithm slightly more the average number than the FACA algorithm. However, the ECARG algorithm outperforms the FACA algorithm in term of accuracy rate by 4.22% and win over 11 from 13 datasets.

ECARG discovers rules with 100% of confidence to build the classifier since the high confidence demonstrates the high possibility of class occurrences when occurring an itemset. Therefore, it gives high accuracy. While the CBA and FACA algorithms build classifier from CARs which passed the minimum confidence threshold. Some of CARs have low confidences so they may predict incorrect class and then the accuracies of CBA and FACA are lower than that of the proposed algorithm.

Table 8. Accuracies of CAB, FACA and ECARG.

Data sets	CBA	FACA	ECARG
Breast cancer	71.52	72.44	**80.00**
Cars	76.85	70.77	**80.92**
Contact-lenses	66.67	68.33	**70.83**
Diabetes	74.47	**73.56**	67.32
Iris	92.67	**95.33**	94
Labor	91.23	91.67	**92.67**
Lymph	78.37	82.43	**88.51**
Mushroom	93.4	96.53	**99.15**
Post-operative	56.67	65.56	**70**
Tic-tac-toe	**98.85**	90.23	95.4
Vote	94.94	91.92	**95.31**
Wined	91.57	92.16	**98.87**
Zoo	80.63	86	**93.07**
Average	82.14	82.40	**86.62**

Table 9. The average number of generated rules on UCI datasets.

Data sets	CBA	FACA	ECARG
Breast cancer	16	24	37
Cars	25	14	22
Contact-lenses	9	5	7
Diabetes	46	18	4
Iris	11	7	4
Labor	14	15	9
Lymph	33	15	19
Mushroom	8	16	22
Post-operative	35	12	12
Tic-tac-toe	20	12	46
Vote	30	12	11
Wined	11	11	9
Zoo	10	11	10
Average	21	14	16

6   Conclusion

In this paper, we proposed an enhanced associative classification method, namely ECARG. The ECARG algorithm avoids a candidate generation by attempting to select a first general rule with the highest accuracy. It easily generates CARs based on vertical data representation and set difference. Moreover, it early reduces a search space to discover a CAR based on rule correlation. These improvements guarantee small classifiers size that doesn't overfit the training dataset and maintain their accuracy rate. For this reason, the proposed algorithm different from the traditional algorithms, it no needs sorting and pruning process. The experiments were conducted on 13 UCI datasets and shows that the ECARG algorithm outperformed the CBA and FACA algorithms in term of accuracy. It can gain a higher accuracy rate than the CBA and FACA algorithms by 4.40% and 5.63% respectively. Our future work, we continue to speed up the ECARG algorithm for finding efficient class association rules.

References

1. Abdelhamid, N.: Multi-label rules for phishing classification. Appl. Comput. Inf. 11(1), 29–46 (2015). https://doi.org/10.1016/j.aci.2014.07.002
2. Abdelhamid, N., Ayesh, A., Thabtah, F.: Phishing detection based associative classification data mining. Expert Syst. Appl. 41(13), 5948–5959 (2014)
3. Abdelhamid, N., Jabbar, A.A., Thabtah, F.: Associative classification common research challenges. In: 2016 45th International Conference on Parallel Processing Workshops (ICPPW), pp. 432–437. IEEE (2016)
4. Agrawal, R., Srikant, R., et al.: Fast algorithms for mining association rules. In: Proceedings of the 20th International Conference on Very Large Data Bases, VLDB, vol. 1215, pp. 487–499 (1994)
5. Alwidian, J., Hammo, B., Obeid, N.: FCBA: fast classification based on association rules algorithm. Int. J. Comput. Sci. Netw. Secur. (IJCSNS) 16(12), 117 (2016)
6. Hadi, W.: ECAR: a new enhanced class association rule. Adv. Comput. Sci. Technol. 8(1), 43–52 (2015)
7. Hadi, W., Aburub, F., Alhawari, S.: A new fast associative classification algorithm for detecting phishing websites. Appl. Soft Comput. 48, 729–734 (2016). https://doi.org/10.1016/j.asoc.2016.08.005
8. Hadi, W., Issa, G., Ishtaiwi, A.: ACPRISM: associative classification based on PRISM algorithm. Inf. Sci. 417, 287–300 (2017). https://doi.org/10.1016/j.ins.2017.07.025
9. Jabbar, M., Deekshatulu, B., Chandra, P.: Heart Disease Prediction System using Associative Classification and Genetic Algorithm. arXiv:1303.5919 [cs, stat], March 2013
10. Li, W., Han, J., Pei, J.: CMAR: accurate and efficient classification based on multiple class-association rules. In: Proceedings 2001 IEEE International Conference on Data Mining, pp. 369–376. IEEE (2001)
11. Liu, B., Yiming Ma, Hsu, W.: Integrating classification and association rule mining. In: Proceedings of the Fourth International Conference on Knowledge Discovery and Data Mining, August 1998
12. Abdelhamid, N., Ayesh, A., Thabtah, F.: Phishing detection based associative classification data mining. Expert Syst. Appl. 41(13), 5948–5959 (2014). https://doi.org/10.1016/j.eswa.2014.03.019. http://www.sciencedirect.com/science/article/pii/S0957417414001481

13. Nguyen, L., Nguyen, N.T.: An improved algorithm for mining class association rules using the difference of Obidsets. Expert Syst. Appl. **42**(9), 4361–4369 (2015). https://doi.org/10.1016/j.eswa.2015.01.002
14. Ogihara, Z.P., Zaki, M., Parthasarathy, S., Ogihara, M., Li, W.: New algorithms for fast discovery of association rules. In: 3rd International Conference on Knowledge Discovery and Data Mining. Citeseer (1997)
15. Quinlan, J.: C4.5: Programs for Machine Learning. Elsevier, Amsterdam (2014)
16. Singh, J., Kamra, A., Singh, H.: Prediction of heart diseases using associative classification. In: 5th International Conference on Wireless Networks and Embedded Systems (WECON), pp. 1–7, October 2016. https://doi.org/10.1109/WECON. 2016.7993480
17. Song, K., Lee, K.: Predictability-based collective class association rule mining. Expert Syst. Appl. **79**(Suppl. C), 1–7 (2017). https://doi.org/10.1016/j.eswa.2017. 02.024. http://www.sciencedirect.com/science/article/pii/S0957417417301069
18. Thabtah, F., Cowling, P., Peng, Y.: MCAR: multi-class classification based on association rule. In: The 3rd ACS/IEEE International Conference on Computer Systems and Applications, January 2005. https://doi.org/10.1109/AICCSA.2005. 1387030
19. Thabtah, F., Hadi, W., Abdelhamid, N., Issa, A.: Prediction phase in associative classification mining. Int. J. Softw. Eng. Knowl. Eng. **21**(06), 855–876 (2011)
20. Wang, D.: Analysis and detection of low quality information in social networks. In: 2014 IEEE 30th International Conference on Data Engineering Workshops, pp. 350–354, March 2014. https://doi.org/10.1109/ICDEW.2014.6818354
21. Zaki, M., Gouda, K.: Fast vertical mining using diffsets. In: Proceedings of the Ninth ACM SIGKDD International Conference on Knowledge Discovery and Data Mining, KDD 2003, pp. 326–335. ACM, New York (2003). https://doi.org/10.1145/ 956750.956788

Children Activity Descriptions from Visual and Textual Associations

Somnuk Phon-Amnuaisuk[1,2(✉)], Ken T. Murata[3], Praphan Pavarangkoon[3], Takamichi Mizuhara[4], and Shiqah Hadi[1,2]

[1] Media Informatics Special Interest Group, CIE, Universiti Teknologi Brunei, Gadong, Brunei
somnuk.phonamnuaisuk@utb.edu.bn, p20191012@student.utb.edu.bn
[2] School of Computing and Informatics, Universiti Teknologi Brunei, Gadong, Brunei
[3] National Institute of Information and Communications Technology, Tokyo, Japan
{ken.murata,praphan}@nict.go.jp
[4] CLEALINKTECHNOLOGY Co., Ltd., Kyoto, Japan
mizuhara@clealink.jp

Abstract. Augmented visual monitoring devices with the ability to describe children's activities, i.e., whether they are asleep, awake, crawling or climbing, open up possibilities for various applications in promoting safety and well being amongst children. We explore children's activity description based on an encoder-decoder framework. The correlations between semantic of the image and its textual description are captured using convolution neural network (CNN) and recurrent neural network (RNN). Encoding semantic information as activation patterns of CNN and decoding textual description using probabilistic language model based on RNN can produce relevant descriptions but often suffer from lack of precision. This is because a probabilistic model generates descriptions based on the frequency of words conditioned by contexts. In this work, we explore the effects of adding contexts such as domain specific images and adding pose information to the encoder-decoder models.

1 Introduction

The ability to identify objects in the scene is not sufficient for describing activities in the scene. The approaches to bridge the gap between object detection and scene understanding are discussed in [1]. The authors outlined two approaches toward bridging the gap between object detection and scene understanding: (i) a knowledge-based framework that infers activities from the agent's actions and (ii) an encoder-decoder framework that infers activities from correlations of actions and content in the scene. The knowledge-based framework attempts to explicitly infer semantic content from the actions of acting agents in the scene [2], while the encoder-decoder framework implicitly infers semantic content by associating textual activity descriptions with visual information in the scene.

Recent encoder-decoder architectures employ *convolutional neural network* (CNN) [3] to encode each image into a feature vector which can be associated

© Springer Nature Switzerland AG 2019
R. Chamchong and K. W. Wong (Eds.): MIWAI 2019, LNAI 11909, pp. 121–132, 2019.
https://doi.org/10.1007/978-3-030-33709-4_11

(i.e., decoded) into appropriate natural language descriptions of a given image using *recurrent neural network* (RNN). The crucial concept in this paradigm is to associate the encoded information of an image to its textual content. The activation output of a hidden CNN layer can be vectorized into a representation that can be associated with the textual descriptions of the scene [4,5]. In this work, new contributions are from the exploration of added domain specific images and the incorporation of key points (or joint positions of human figures) to augment the visual information encoding of an image. The *long short term memory* (LSTM) [6] is employed to decode the visual information into natural language descriptions.

This paper reports our work in progress from the ASEAN-IVO project titled *Event Analysis: Applications of computer vision and AI in smart tourism industry*. Here, we explore the generation of scene semantics from visual and textual associations using CNN and LSTM. The findings can be applied in extracting museum visitors' profiles and other behavioral studies. The rest of the materials in this paper are organized into the following sections; Sect. 2 describes related works; Sect. 3 highlights the technicality of our approach; Sect. 4 describes the experimental designs and presents experimental results; and Sect. 5 provides discussions, and future direction.

2   Related Works

Computer vision before the deep learning era relied heavily on the concept of discriminative features. The traditional feature-based approach engineers features from the pixel information of an image [7]. *Color*, and *intensity* are the primitives where other structures can be derived from them, e.g., *edge*, *contour*, *color histogram*, and *intensity gradient*. Intensity gradient peaks are more robust to illumination variations since they have good repeatability and provide good descriptions of local appearances described as points and shapes. Many elaborated representation structures such as *Kanade-Lucas-Tomasi (KLT)* corner detectors [8], *Scale Invariant Feature Transform (SIFT)* [9], *Speeded Up Robust Features (SURF)* [10] and *Histogram of Gradient (HOG)* prove to be expressive hand crafted features and have been exploited to represent appearances of humans and other objects in the scene.

The *ImageNet* [11] introduces a large scale image classification depository to the computer vision research community. Various deep learning models have been explored and have shown significant improvements in classification accuracy over the hand crafted feature-based approach in the ImageNet *Large Scale Visual Recognition Challenge* (ILSVRC) competition. Notably, the following variant of CNN architectures have been explored: *AlexNet* [12], *GoogLeNet* [13], *VGG* [14], *ResNet* [15] and *MobileNet* [16]. In essence, the CNN is characterized by groups of convolution and pooling layers before connecting to the fully connected layers at the output-end. The CNN can be trained to learn effective features from a large amount of training images in an end-to-end manner (i.e., the network accepts the input image and outputs the desired classification label). The aforementioned various networks explored various variations in terms of the number and size of convolution filters, the number of convolution layer and the way of the convolution groups are connected together, e.g., the Inception unit in GoogLenet exploits various filter sizes and the ResNet has a skip layer.

The CNN has been successfully employed as a recognizer of a single image as a whole unit. With an added region proposal mechanism [17], many object instances can be located in a single input image. As a matter of facts, the existence of different objects in the visual scene implicitly captures the semantic of the scene in CNN. For example, a scene with sky, sea gal and sand is likely to be an outdoor scene, perhaps a beach scene; on the other hand, a scene with many bottles, glasses and people is likely to be an indoor scene, perhaps a pub or a kitchen.

In [4], the authors present a model that generates natural language descriptions of the objects and scene in respective images. The model combines both CNN and RNN, and the authors have experimented using the Flickr8k, Flickr30k and MSCOCO2014 data sets. In brief, a vector v representing contents of various visual regions in an image is extracted using CNN models. The vector $V = [v_1, ..., v_i]$ is employed to train a Recurrent Neural Network (RNN) model to generate natural language descriptions based on the textual description of the image. The model in [4] can successfully generate natural language descriptions $D = [d_1, ..., d_j]$ of an input image. Although the semantic is not extracted directly from the image, but rather as an association between a vector V and generated descriptions D, the results show that implicit associations between contents in an image and its descriptions in the natural language are feasible. We are inspired by this work and decided to explore the approach to describe children's activities in an image.

Apart from the object-activity correlation via the deep learning tactic discussed above, many other related works have been explored in associating sensor information to activities. For examples, a knowledge-based approach explicitly infers semantic contents in the scene [18,19] and a data mining approach explores multi signal motifs mining [20,21].

3 Encoding-Decoding Children Activities

When feeding an input image to a CNN, the input image is transformed into various representations via convolution operations, pooling operations and feed-forward operations. These representations can be retrieved from CNN layers. The representation at the output layer signifies the object class. The representation at other hidden layers can be interpreted as various encoded representations of the input image.

Let (A, D) be a tuple of an image $A^{r \times c}$ and its textual description D, let cnn be an instance of a CNN class trained with ImageNet data. The CNN is trained using supervised learning and it can learn appropriate connection weights among layers, i.e., convolution filters $w^{k_r \times k_c}$, and connection weights of the feed-forward layers. The output v^l can be retrieved from the activation of neurons in layer l:

$$v_{x,y}^l = h(\sum_{k_r} \sum_{k_c} w^{k_r \times k_c} a_{x:x+k_r, y:y+k_c} + b) \tag{1}$$

where $h(\cdot)$ is an activation functions, e.g., ReLU, sigmoid, tanh, $a^{k_r \times k_c}$ is a set of input patches extracted by windowing through the 2D layer A, b, which is a bias and $v_{x,y} \in R$ is an output at the next layer l. The head of CNN is a typical feed-forward neural network and the weights (i.e., convolution filters and fully connected layers) of CNN are learned using the standard backward propagation technique from the loss L.

$$\frac{dL}{dW} = \frac{dL}{dV^1} \cdots \frac{dV^l}{dW} \tag{2}$$

After the CNN network has learnt the domain, the weight of the network is frozen. The *cnn* takes an input image $cnn.input(I)$ and a vector V can be retrieved from layer L^l of the *cnn* object, $V^l = cnn.L^l$. The vector V above encodes the visual information of the image I, since D is a textual description of I. Information from both V and D can be used to describe the scene. The LSTM model is a recurrent neural network for this decoding task. In the visual description task, this can be accomplished by learning to associate a sequence of textual input $S = [s_1, ..., s_n]$ to the textual output $S' = s_{n+1}$ conditioned by the visual information V i.e., $p(S'|S, V)$.

Figure 1 below shows the concept of visual-textual information encoding, how they are combined and decoded to generate natural language description of an image. In brief, each word s in the textual description D is tokenized and transformed into embedded representations s^g based on global vectors for word representation (GloVe) [22]. This process encodes each word into a vector of real numbers. The input V and s^g represent visual and textual encoded representations, respectively.

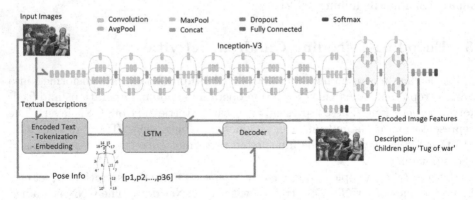

Fig. 1. The overall concept of the visual-textual association. Visual information is encoded in the form of (i) output from Inception-V3 and (ii) pose information extracted using pose-estimation techniques. The LSTM is trained to generate words conditioned by visual information.

Table 1. Summary of number of trained samples employed in our visual-textual association models. Three experiments are carried out, each using three data sets. There are a total of nine models.

	Flickr8k	Flickr30k	MSCOSO 2017	Others
Exp 1: Trained visual-textual pairs	30,000	127,000	150,000	
Exp 2^{α}: Trained visual-textual pairs	30,000	127,000	150,000	500
Exp 3^{β}: Trained visual-text pairs	30,000	127,000	150,000	Pose info

α: 500 domain specific visual-text pairs are included in the training data
β: Pose information is included in the visual information

3.1   Visual and Textual Information

Fortunately, the existing *Flickr8k, Flickr30k* and *MSCOCO 2017* data sets provide an excellent resource for images and their corresponding description in natural language. Each image in these data sets is described with at least five descriptions. Those images capture general visual scenes in everyday life and are not specific to children's activity. In this work, we create another 100 domain specific images of children in various activities collected from the Internet and Youtube[1]. These domain specific images, where each image is also described with five textual descriptions, will be included in the training process. Figure 2 shows examples randomly pulled out images from the data sets. Relevant information about the construction of our visual-textual association models from these images is summarized in Table 1.

Encoding Pixel Information. The pixel information is encoded using the Inception V3 model pre-trained with the ImageNet dataset[2]. Each input image is transformed into an encoded vector V^l, ($|V^l| = 2048$) which is the output from the *Global Average Pooling* layer. Three experiments were carried out. In the first experiment, three models were built using data from Flickr8k, Flickr30k and MSCOCO2017. We employed around 80–85% of the available data in Flickr8k and Flick30k to train the model. The MSCOCO2017 offers a large amount of images (over 118,000 images) and only 30,000 images (i.e., 150,000 visual-text pairs) were employed to train our model. This decision is from our own limitation in computing resources. In the second experimental design, 500 domain specific visual-text pairs were added into the training data for all models. This aims to investigate the influence of domain specific description inputs to the generated descriptions. In the third experiment, pose informations were included as an additional source of information.

---

[1] Youtube standard license for fair use of public content.
[2] www.image-net.org/challenges/LSVRC.

Fig. 2. Random examples of images and descriptions from Row 1: Flickr8k dataset, Row 2: Flickr30k dataset, Row 3: MSCOCO 2017 dataset and Row 4: Our own domain specific dataset

Encoding Pose Information. The human pose can clearly reveal information regarding body and limbs positions. Hence pose information can provide expressive information regarding a person's action, which should implicitly correlate with descriptions of the scene.

Each pose holds a sequence of 18 joint positions in a 2D space, $P = [p_1, ..., p_{18}]$ where p_i denotes (x_i, y_i) coordinate information of a key point. We decided to include two sets of pose information. The pose data in the first set is normalized by projecting points in P in terms of image width and height and the pose data in the second set is normalized by projecting (x_i, y_i) according to $[min(x_i), max(x_i)]$ and $[min(y_i), max(y_i)]$ scales. Hence the final P is a vector of $[p_1, ..., p_{36}]$ where $(x_i, y_i) \in [0, 1]$.

In this work, we compute pose information using https://github.com/ildoonet/tf-pose-estimation. All training images from Flickr8k, Flickr30k and MSCOCO2017 are run through the algorithm and one pose is retrieved from each image. In the case that no pose is detected in an image, all entries of the pose vector are set to zeros.

Encoding-Decoding Textual Information. Each image is accompanied with five descriptions (see examples in Fig. 2). All descriptions in the whole training set are processed and each sentence can be described as a vector of vocabulary. Each vocabulary vector represents a single word and it will be transformed into an embedded vector s^g before feeding to the LSTM.

Figure 3 (left pane) illustrates an LSTM unit. At time t, input X is the concatenation of $X_t|H_{t-1}$ where X_t is the input at time t and H_{t-1} is the delayed input from the previous time step. The flow of information is as follows: $H_{t-1} \Rightarrow H_t$; $C_{t-1} \Rightarrow C_t$ and $X_t \Rightarrow Y_t$. These are expressed in the following equations:

$$f_t = \sigma(W_f[X_t, H_{t-1}] + b_f) \tag{3}$$

$$i_t = \sigma(W_i[X_t, H_{t-1}] + b_i) \tag{4}$$

$$o_t = \sigma(W_o[X_t, H_{t-1}] + b_o) \tag{5}$$

$$C'_t = tanh(W_c[X_t, H_{t-1}] + b_c) \tag{6}$$

$$C_t = (f_t \otimes C_{t-1}) \oplus (i_t \otimes C'_t) \tag{7}$$

$$H_t = o_t \otimes tanh(C_t) \tag{8}$$

$$Y_t = softmax(W H_t + b) \tag{9}$$

where the operator $\otimes$ denotes the Hadamard, element-wise product. Matrices W_g denotes the weights, where g can either be the input gate i, output gate o, the forget gate f or the memory cell c. The functions $\sigma, tanh$ and $softmax$ denote the sigmoid function, the hyperbolic tangent function and $softmax$ function, respectively.

We present the abstraction of the encoding-decoding process implemented in our experiments in Fig. 3. For a given image and description training data, the LSTM is initialized with encoded information from the image (the initial values of C_0 and H_0) and is trained with input-out word pairs derived from the wearable sensor-based approach that associates movement to activity (see Fig. 3 middle pane). In other words, feeding sequence of words $[s_{t-n}, ..., s_t]$ to LSTM, it predicts output word s_{t+1} according to the probability $p(s_{t+1}|s_{t-n}, ..., s_t, V)$, where V is the visual context.

Pose information is added in the third experiment, here, it is decided to concatenate the pose vector to the output vector of LSTM before the decoding stage (see Fig. 3 right pane).

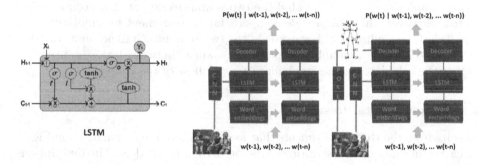

Fig. 3. Left pane: a schematic diagram of the LSTM; Middle pane: the conceptual diagram of the setup in our experiment one and two; Right pane: the conceptual diagram of the setup in the experiment three. Note that the encoded visual information from pixels is used to initialize the LSTM, while joint information extracted from the image is used in conjunction with the output from LSTM.

4   Experimental Setup and Discussion

We employed 3000 images from the MSCOCO2017 validation sets for the evaluation of all models. These images are not seen by the models constructed from the training data sets of Flickr8k, Flickr30k, and MSCOCO2017. Figure 4

two zebras are standing in the grass

man holding cell phone in his hand

man is using laptop computer

black dog is running through the grass

Fig. 4. Examples of generated descriptions from our experiments

shows some representative generated descriptions. In these specific examples, the description "two zebras are standing in the grass" seems appropriate. The other two descriptions "man holding cell phone in his hand" and "man is using laptop computer" are not exactly right but they are understandable. However, "black dog is running through the grass" does not make sense. Upon examining these descriptions, it is clear that designing objective evaluation criteria to evaluate the semantic of the content is a great challenge.

Evaluation Criteria. In order to provide an objective justification on the quality of the output, we calculate the semantic similarities between the generated description and the pre-defined descriptions of the unseen test images. Similarities between two sentences are calculated by converting each sentence into a vector of word occurrences. Therefore, two sentences sharing all key words yield *cosine similarity* of 1 and yield 0 if the two sentences do not share common key words. However, this can provide a superficial measurement for similarity as it only takes the similarity of key words in the two sentences into account. In other words, the higher the number of shared key words between two sentences, the higher the calculated value for similarity, regardless of the context.

4.1 Experimental Results

Results from the three experiments that we have set up are summarized in Fig. 5. Each picture has nine descriptions generated by nine models. The descriptions were generated in the order of experiment 1, 2 and 3 with each experiment providing three descriptions. The three models in each experiment were built from the datasets Flickr8k, Flickr30k and MSCOCO17. Experiment 1 learnt visual-textual correlations from CNN-LSTM architecture to generate three descriptions. In experiment 2, domain specific images were added to the training process and generates three other descriptions. Experiment 3 utilises pose information in its training process to provide three final descriptions. Each description is preceded with an average similarity score. Similarity scores between the generated description and the five descriptions predefined in the dataset are used to compute the average score.

0.136 little boy is playing with toy
0.192 boy in blue shirt is holding the arm of the adult who is holding baby
0.092 young boy is playing with batman
0.129 man in white shirt and black shorts plays
0.055 man in blue shirt is holding baby in blue shirt
0.168 man in blue shirt is sitting on the floor with his hands on his lap
0.066 man in black shirt holding cell phone
0.086 toddler is crawling in the sand
0.155 man is sitting on couch with his laptop

0.223 woman in black and white dress is standing by an umbrella
0.106 woman in black and black dress is sitting on red bench
0.098 two women sit on the sidewalk under an umbrella
0.127 two women are sitting on bench in front of an old building
0.236 man in white shirt is standing in front of an old building
0.221 man in blue shirt and jeans is standing in front of an atm
0.268 woman holding an umbrella while standing in front of crowd
0.137 child is holding hot dog
0.155 man walking down street holding an umbrella

0.139 young boy is sitting on bed with his hands up
0.240 boy in blue shirt is holding the arm of the girl who is holding baby
0.098 little boy in blue shirt is playing with his toy
0.401 young girl is brushing her teeth
0.153 young boy in blue shirt is holding toy
0.133 young boy is playing with toy in bathtub
0.114 man is holding donut in his hand
0.082 child is crawling in sandbox
0.114 man is holding donut in his hand

0.076 two boys are playing with water
0.153 two children are playing in the grass
0.166 two boys are playing on the grass
0.276 two children are sitting on the grass
0.225 man in blue shirt is sitting on the ground
0.240 young girl in pink shirt is walking on the sidewalk
0.067 man in black shirt and jeans is holding an umbrella
0.043 child is crawling in sandbox
0.207 man and woman are sitting on bench

0.231 group of people are standing around at night
0.188 group of people are gathered around carnival game
0.198 group of kids are playing with water balloons
0.346 group of people are sitting around table with drinks
0.274 group of people are sitting in front of large crowd of people
0.329 group of people are sitting at table with drinks and drinks
0.237 man in suit and tie sitting at table
0.177 group of people are playing in sandbox
0.345 group of people sitting around table with laptops

0.183 man in white shirt and jeans playing tennis
0.144 tennis player hitting the ball
0.129 man in white shirt and black shorts plays tennis
0.122 tennis player returns to hit the ball
0.196 man in white shirt is playing tennis
0.077 two men playing basketball one is hitting the ball and the other is wearing white and white
0.221 man is playing tennis on the court
0.194 man is playing tennis
0.283 man holding tennis racquet on tennis court

Fig. 5. Examples of images, each with nine descriptions generated from nine models.

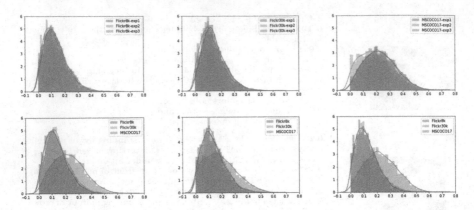

Fig. 6. Top row: comparing density plots of models trained using different context; Bottom row: comparing density plots from models trained using different dataset but with the same context, e.g., added pose information.

4.2   Discussion

The study shows that the models trained using MSCOCO 2017 dataset show better semantic similarity scores, compared to the models trained using Flickr8k and Flickr30k (see density plots in Fig. 6 bottom row). The effects of adding extra context in terms of extra images and pose information are, however, inconclusive and deserve further investigation (see density plots in Fig. 6 top row).

From previous works in image descriptions generation [4,5], it is known that the model can sometimes generate sensible descriptions that humans could agree on, but most of the time, it generates understandable descriptions but irrelevant to the context; occasionally it generates completely inaccurate descriptions. This is still an open research issue and we believe the deficiency is from the fact that the generated text is confabulated based on the visual-textual correlations. Here, the second and third experiments were carried out to enhance this point. However, in this experiment, adding domain specific information and adding extra pose information did not show a clear improvement of the quality of the descriptions.

Human activities are hierarchically organized and can be hierarchically described at different grain sizes. For example, a scene may be described as *a father teaches his son to kick a ball* or *a boy kicks a ball while a man is watching*. Also the same scene can be described using primitive actions of limb movements of each actor and object involved in the activity; these different descriptions are suitable for different applications. This is a hard problem and we do not know an effective algorithm for this problem yet.

In our opinion, the encoder-decoder paradigm could be applicable in domains where general descriptions of the scene (where specificity is not a priority) are sufficient, for example, describing traffic as heavy or light, describing weather as rainy, sunny or cloudy, describing people in a scene without detailed profiles. However, a knowledge based approach is more suitable in providing detailed descriptions as it is more flexible and natural to incorporate additional knowledge and analytical functionalities to the scene, e.g., analyse movements, strategy in a football match, boxing and other skill based activities.

5   Conclusion and Future Directions

The ability to describe children's activities from a visual scene automatically is important in many domains, e.g., early childhood development, smart nursery, smart tourism. This work investigates children's activity descriptions from visual and textual associations. The computational model is built from recent advances in deep learning. Visual information is derived from the *Global Average Pooling* layer of the Inception-V3 network. This work explores the encoder-decoder paradigm and further explore the influence of adding visual information in terms of human joint positions (skeleton key points) as well as adding extra domain specific information into the training data.

In future works, the encoder-decoder approach can be enhanced with (i) temporal information by encoding information from a sequence of images and (ii) space-time interest point features or action related information such as correlation among pose information and other objects in the scene.

Acknowledgments. This publication is the output of the ASEAN IVO http://www. nict.go.jp/en/asean_ivo/index.html project titled *Event Analysis: Applications of computer vision and AI in smart tourism industry* and financially supported by NICT (http://www.nict.go.jp/en/index.html). We wish to thank Centre for Innovative Engineering (CIE), Universiti Teknologi Brunei for their partial financial support given to this research. We would also like to thank anonymous reviewers for their constructive comments and suggestions.

References

1. Phon-Amnuaisuk, S., Murata, K.T., Pavarangkoon, P., Mizuhara, T., Yamamoto, K., Mizuhara, T.: Exploring the applications of faster R-CNN and single-shot multi-box detection in a smart nursery domain. arXiv:1808.08675 (2018)
2. Kojima, A., Tamura, T., Fukunaga, K.: Natural language description of human activities from video images based on concept hierarchical of actions. Int. J. Comput. Vision **50**(2), 171–184 (2002)
3. LeCun, Y., Bengio, Y., Hinton, G.: Deep learning. Nature **521**, 436–444 (2015)
4. Karpathy, A., Fei-Fei, L.: Deep visual-semantic alignments for generating image descriptions. In: Proceedings of the International Conference on Computer Vision and Pattern Recognition (CVPR). CoRR abs/1412.2306 (2015)
5. Vinyals, O., Toshev, A., Bengio, S., Erhan, D.: Show and tell: lessons learned from the 2015 MSCOCO image captioning challenge. IEEE Trans. Pattern Anal. Mach. Intell. **39**(4), 652–663 (2016)
6. Hochreiter, S., Schmidhuber, J.: Long short-term memory. Neural Comput. **9**(8), 1735–1780 (1997)
7. Gonzalez, R.C., Woods, R.E.: Digital Image Processing. Prentice Hall, Upper Saddle River (2001)

8. Shi, J., Tomasi, C.: Good features to track. In: Proceedings of the IEEE Conference on Computer Vision and Pattern Recognition, pp. 593–600 (1994)
9. Lowe, D.G.: Object recognition from local scale-invariant features. In: Proceedings of the Seventh IEEE International Conference on Computer Vision, Kerkyra, vol. 2, pp. 1150–1157 (1999)
10. Bay, H., Ess, A., Tuytelaars, T., Van Gool, L.: SURF: speeded up robust features. Comput. Vis. Image Underst. (CVIU) **110**(3), 346–359 (2008)
11. Deng, J., Dong, W., Socher, R., Li, L.-J., Li, K., Fei-Fei, L.: ImageNet: a large-scale hierarchical image database. In: IEEE Conference on Computer Vision and Pattern Recognition (CVPR), pp. 248–255 (2009)
12. Krizhevsky, A., Sutskever, I., Hinton, G.E.: Imagenet classification with deep convolutional neural networks. In: Proceedings of the 25th International Conference on Neural Information Processing Systems (NIPS), pp. 1097–1105 (2012)
13. Szegedy, C., et al.: Going deeper with convolutions. In: Proceedings of the International Conference on Computer Vision and Pattern Recognition (CVPR) (2015)
14. Simonyan, K., Zisserman, A.: Very deep convolutional networks for large-scale image recognition. In: Proceedings of the International Conference on Learning representations (ICLR) CoRR arXiv: 1409.1556 (2015)
15. He. K., Zhang, X., Ren, S., Sun, J.: Deep Residual Learning for Image Recognition. CoRR, abs/1512.03385 (2015). http://arxiv.org/abs/1512.03385
16. Howard, A.G., et al.: MobileNets: efficient convolutional neural networks for mobile vision applications. CoRR, abs/1704.04861 (2015). http://arxiv.org/abs/1512.03385
17. Girshick, R.: Fast R-CNN. In: Proceedings of the IEEE Conference on Computer Vision (ICCV), pp. 1140–1148 CoRR arXiv:1504.08083v2 (2015)
18. Ye, J., Stevenson, G., Dobson, S.: USMART: an unsupervised semantic mining activity recognition technique. ACM Trans. Interact. Intell. Syst. (TiiS) **4**(4), 16 (2015)
19. Civitarese, G., Bettini, C., Sztyler, T., Riboni, D., Stuckenschmidt, H.: newNEC-TAR: collaborative active learning for knowledge-based probabilistic activity recognition. Pervasive Mob. Comput. **56**(2019), 88–105 (2019)
20. Vahdatpour, A., Amini, N., Sarrafzadeh, M.: Toward unsupervised activity discovery using multi-dimensional motif detection in time series. In: Proceedings of the 21st International Joint Conference on Artificial Intelligence (IJCAI 2009), pp. 1261–1266 (2009)
21. Rashidi, P., Cook, D.J., Holder, L.B., Schmitter-Edgecombe, M.: Discovering activities to recognize and track in a smart environment. IEEE Trans. Knowl. Data Eng. **23**(4), 527–539 (2011)
22. Pennington, J., Socher, R., Manning, C.D.: GloVe: global vectors for word representation. In: Empirical Methods in Natural Language Processing (EMNLP), pp. 1532–1543 (2014). http://www.aclweb.org/anthology/D14-1162

Randomspace-Based Fuzzy C-Means for Topic Detection on Indonesia Online News

Muhammad Rifky Yusdiansyah$^{(\boxtimes)}$ ⓘ, Hendri Murfi ⓘ,
and Arie Wibowo ⓘ

Department of Mathematics, Universitas Indonesia, Depok 16424, Indonesia
{muhammad.rifky51,hendri,arie.wibowo}@sci.ui.ac.id

Abstract. Topic detection is a process used to analyze words in a collection of textual data to determine the topics in the collection, how they relate to each other, and how they change from time to time. Fuzzy C-Means (FCM) and Kernel-based Fuzzy C-Means (KFCM) method are clustering method that is often used in topic detection problems. Both FCM and KFCM can group dataset into multiple clusters on a low-dimensional dataset, but fail on high-dimensional dataset. To overcome this problem, dimension reduction is carried out on the dataset before topic detection is carried out using the FCM or KFCM method. In this study, the national news account's tweets dataset on Twitter were used for topic detection using the Randomspace-based Fuzzy C-Means (RFCM) method and Kernelized Randomspace-based Fuzzy C-Means (KRFCM) method. The RFCM and KRFCM learning methods are divided into two steps, which are reducing the dimension of the dataset into a lower-dimensional dataset using random projection and conducting the FCM learning method on the RFCM and the KFCM learning method on KRFCM. After obtaining the topics, then an evaluation is carried out by calculating the coherence value on the topics. The coherence value used in this study uses the Pointwise Mutual Information (PMI) unit. The study was conducted by comparing the average PMI values of RFCM and KRFCM with Eigenspace-based Fuzzy C-Means (EFCM) and Kernelized Eigenspace-based Fuzzy C-Means (KRFCM). The results obtained using national news account's tweets showed that the RFCM and KRFCM methods offered faster running time for a dimensional reduction but had smaller average PMI values compared to the average PMI values generated by the EFCM and KEFCM learning methods.

Keywords: Topic detection · Random projection · Fuzzy C-Means · Twitter

1 Introduction

Information and communication technology is overgrowing; this is characterized by the ease of obtaining information through the Internet. According to the data from the Internet Live Stats site, internet users in the world currently reach four billion, and the number continues to grow over time [1]. The rapid development of Internet technology is also followed by increasing the amount of information. Information can be obtained through social media, one of which is Twitter. Twitter allows the users to send messages, known as a tweet, about real-world events almost in real-time. More than

© Springer Nature Switzerland AG 2019
R. Chamchong and K. W. Wong (Eds.): MIWAI 2019, LNAI 11909, pp. 133–143, 2019.
https://doi.org/10.1007/978-3-030-33709-4_12

29 billion tweets are on Twitter and growing up to 300 million every day [2]. It makes Twitter becomes real-time information resource about real-world events locally in a region or globally in the world [3].

One of the information available on Twitter is a topic that is currently being discussed. To gather that information, we need some process called topic detection. Topic detection analyzes words in a collection of documents so that the topics contained in the document collection are found, how the topics relate to each other, and how the topic changes over time [4]. Topic detection can be done manually by reading all the textual data. However, due to a massive number of tweets, this manual way is difficult or even impossible. Therefore, a machine learning approach is needed to detect the topics.

One of topic detection methods is a clustering-based method. In the clustering method, the textual data is grouped in a way that members of the same cluster have more similarity to each other than with other group members. The centers of the clusters, called centroids, are means of their members. In topic detection problem, the centroids are interpreted as topics. K-means is a standard clustering method for topic detection [5–7]. This method splits the textual data into k clusters in which each textual data belongs to the nearest cluster center. In other words, the k-means method assumes that each textual data only belongs to one topic. This assumption is rather weak because most of the textual data have more than one topic. Therefore, soft clustering approach to be considered where each textual data may belong to more than one clusters. One of the soft clustering methods is fuzzy c-means (FCM) [8]. There is a modification of FCM using kernel trick called kernel-based fuzzy c-means (KFCM) [9]. Both FCM and KFCM works well on low dimensional data but fails on high dimensional data [10]. To overcome this problem, a method called Eigenspace-based Fuzzy C-Means (EFCM) and Kernelized Eigenspace-based Fuzzy C-Means (KEFCM) are proposed by transforming the high dimensional data into a lower-dimensional space using singular value decomposition and both methods are performed in this lower-dimensional space [11–13].

In this paper, we examine another projection method called random projection to transform the high dimensional data into lower-dimensional space. We called this method as randomspace-based fuzzy c-means (RFCM) and kernelized randomspace-based fuzzy c-means (KRFCM). We examine the performance of those methods for topic detection on Indonesian online news from Twitter. Our simulation shows that both RFCM and KRFCM achieve lower performance than both EFCM and KEFCM regarding topic interpretability in term of topic coherence. However, both RFCM and KRFCM are much more efficient and scalable.

The rest of the paper is organized as follows: Sect. 2 we present the reviews FCM, KFCM, GRP, RFCM, and KRFCM. Section 3 describes the results of our simulations and discussions. Finally, a general conclusion about the results is presented in Sect. 4.

2 Methods

Given textual data, the processes of clustering-based topic detection start by using the fuzzy c-means algorithm (FCM) [8] (Sect. 2.1) or kernel-based fuzzy c-means algorithm (KFCM) [9] (Sect. 2.2). Since textual data usually have high dimension and both FCM and KFCM fail in clustering the high dimensional data [10] Therefore, we use random projection (Sect. 2.3) to transform the textual data into the lower-dimensional space. We called this clustering-based topic detection algorithm as randomspace-based fuzzy c-means (RFCM) (Sect. 2.4) and kernelized randomspace-based fuzzy c-means (KRFCM) (Sect. 2.5).

2.1 Fuzzy C-Means

Fuzzy C-Means (FCM) produce some clusters by divide dataset into membership values $R = [r_{nk}]$, with r_{nk} values are between 0 and 1. The r_{nk} values implicitly say that one data can belong to one or more clusters. This method based on the minimization of the objective function as expressed:

$$J_{FCM}(R, M) = \sum_{k=1}^{K} \sum_{n=1}^{N} r_{nk}^{w} \|x_n - \mu_k\|^2 \tag{1}$$

where $\|\cdot\|$ is any norm represent similarities between the centroid and data, and w is a fuzzification constant to decide the level of fuzziness.

As the objective function is minimized, values in high membership area are given to data near the cluster centroid. The membership values and the centroids are computed iteratively using alternating optimization. Algorithm 1 below is the algorithm of fuzzy c-means.

Algorithm 1. Fuzzy C-Means [8]

Input   : X, w, max number of iterations (T), threshold (ε)
Output : r_{nk}, μ_k
1.   set $t = 0$
2.   initialize μ_k
3.   update $t = t + 1$

4.   calculate $r_{nk} = \left[\sum_{j=1}^{K} \left(\frac{\|x_n - v_k\|_2}{\|x_n - v_j\|_2} \right)^{2/w-1} \right]^{-1}, \forall n, k$

5.   calculate $\mu_k = \frac{\sum_{n=1}^{N} ((r_{nk})^w x_n)}{\sum_{n=1}^{N} (r_{nk})^w}, \forall k$

6.   if a stopping, i.e., $t > T$ or $\|R^t - R^{t-1}\|_F < \varepsilon$, is fulfilled then stop, else go back to step 3

2.2   Kernel-Based Fuzzy C-Means

Kernel-based Fuzzy C-Means (KFCM) works the same way as FCM but uses kernel function to measure the similarity between data and cluster centroid. The kernel function is used to make the clustering process performed in higher dimensional space without explicitly transforming the textual data.

Define nonlinear map as $\phi : x \rightarrow \phi(x) \in F$, where $x \in X$. X denotes the data space, and F the transformed feature space with a higher or even infinite dimension. Kernel function can be express as $ker(x, y) = \phi(x) \cdot \phi(y)$ which represents the inner product of ϕ function. Kernelized FCM minimizes the following objective function.

$$J_{KFCM}(R, \ M) = \sum\nolimits_{k=1}^{K} \sum\nolimits_{n=1}^{N} r_{nk}^{w} \|\phi(x_n) - \phi(\mu_k)\|^2 \qquad (2)$$

where

$$\|\phi(x_n) - \phi(\mu_k)\|^2 = ker(x_n, x_n) + ker(\mu_k, \mu_k) - 2ker(x_n, \mu_k) \qquad (3)$$

In simulations, we use RBF kernel functions, i.e., $ker(x, y) = exp\left(-\gamma\|x - y\|^2\right)$, so $ker(x, x) = 1$. According to Eq. 3, Eq. 2 can be written as

$$J_{KFCM}(R, \ M) = 2 \sum\nolimits_{k=1}^{K} \sum\nolimits_{n=1}^{N} r_{nk}^{w}(1 - ker(x_n, \mu_k)) \qquad (4)$$

Minimizing Eq. 4 using alternating optimization under constraining M, we have

$$r_{nk} = \left(\frac{1}{\sum_{j=1}^{K} \left(\frac{1 - ker(x_n, \mu_k)}{1 - ker(x_n, \mu_j)} \right)} \right)^{\frac{1}{w-1}} \qquad (5)$$

$$\mu_k = \frac{\sum_{n=1}^{N} r_{nk}^{w} ker(x_n, \mu_k) x_n}{\sum_{n=1}^{N} r_{nk}^{w} ker(x_n, \mu_k)} \qquad (6)$$

As the objective function is minimized, values in high membership area are given to data near the cluster centroid. The membership values and the centroids are computed iteratively using alternating optimization. Algorithm 2 below is the algorithm of kernel-based fuzzy c-means.

Algorithm 2. Kernel-based Fuzzy C-Means [9]

Input   : X, w, max number of iterations (T), threshold (ε)

Output : r_{nk}, μ_k

1. set $t = 0$
2. initialize μ_k
3. update $t = t + 1$
4. calculate membership matrix using Eq. (5)
5. calculate cluster centroid using Eq. (6)
6. if a stopping, i.e., $t > T$ or $\|R^t - R^{t-1}\|_F < \varepsilon$, is fulfilled then stop, else go back to step 3

2.3 Random Projection

Random Projection is one of the reduction dimension methods for the matrix which has many uses in the data processing. There is no formal definition of Random Projection, but we will give the Definition 1 according to [14].

Definition 1. A random linear map $T : \mathbb{R}^m \to \mathbb{R}^k$ is called a random projection (or random matrix) if for all $\varepsilon \in (0, 1)$ and all vectors $x \in \mathbb{R}^m$, we have:

$$\mathbb{P}\left((1 - \varepsilon)\|x\|^2 \leq \|T(x)\|^2 \leq (1 + \varepsilon)\|x\|^2 \right) \geq 1 - 2e^{-C\varepsilon^2 k} \tag{7}$$

For some universal constant $C > 0$ (independent of m, k, ε).

The random projection itself based upon the Johnson-Lindenstrauss lemma proposed in 1984 which state that "*A set of points in a high-dimensional space can be projected into a lower-dimensional subspace of in such a way that relative distances between data points are nearly preserved*" [15]. Note that the projected lower dimension subspace is selected randomly based on some distribution. In this case, we use Gaussian distribution. Algorithm 3 below is the algorithm of random projection.

Algorithm 3. Random Projection

Input   : X, data dimension (d), reduced dimension (c)

Output: $\tilde{X}$

1. initialize random matrix $R_{c \times d}$ using $Gaussian(0, \frac{1}{c})$ distribution
2. calculate $\tilde{X}$ from multiplying $R_{c \times d}$ and X
3. end

2.4   Randomspace-Based Fuzzy C-Means

Simply randomspace-based fuzzy c-means (RFCM) is to use random projection before using FCM. Random projection reduces the dimension of data X, as described in Algorithm 3. As mentioned above, the output of random projection is reduced matrix $\tilde{X}$.

Next, we perform FCM using matrix on the reduced dimensional data $\tilde{X}$. In this step, we calculate the membership matrix R based on $\tilde{X}$. To calculate the centroid matrix, we multiply the membership matrix R with the original data X. Algorithm 4 below is the algorithm of randomspace-based fuzzy c-means.

Algorithm 4. Randomspace-based Fuzzy C-Means

Input   : X, w, data dimension (d), reduced dimension (c), max number of
              iterations (T), threshold (ε)

Output : r_{nk}, μ_k
1.   set $t = 0$
2.   reduce the dimension of the data X to c-dimensional data $\tilde{X}$ using random projection (Algorithm 3)
3.   initialize μ_k
4.   update $t = t + 1$
5.   calculate $r_{nk} = \left[\sum_{j=1}^{K} \left(\frac{\|x_n - v_k\|_2}{\|x_n - v_j\|_2} \right)^{2/w-1} \right]^{-1}, \forall n, k$
6.   calculate $\mu_k = \frac{\sum_{n=1}^{N} ((r_{nk})^w x_n)}{\sum_{n=1}^{N} (r_{nk})^w}, \forall k$
7.   if a stopping, i.e., $t > T$ or $\|R^t - R^{t-1}\|_F < \varepsilon$, is fulfilled then stop, else go back to step 4
8.   calculate cluster centroid from original data with multiplying R and original data X

2.5   Kernelized Randomspace-Based Fuzzy C-Means

Same as RFCM, kernelized randomspace-based fuzzy c-means (KRFCM) is to use random projection before using KFCM. Random projection reduces the dimension of data X, as described in Algorithm 3. As mentioned above, the output of random projection X is reduced matrix $\tilde{X}$.

Next, we perform FCM using matrix on the reduced dimensional data $\tilde{X}$. In this step, we calculate the membership matrix R based on $\tilde{X}$. To calculate the centroid matrix, we multiply the membership matrix R with the original data X.

Algorithm 5. Kernelized Randomspace-based Fuzzy C-Means	

Input   : X, w, data dimension (d), reduced dimension (c), max number of
         iterations (T), threshold (ε)
Output : r_{nk}, μ_k
  1.   set $t = 0$
  2.   reduce the dimension of the data X to c-dimensional data $\tilde{X}$ using random-
       dom projection (Algorithm 3)
  3.   initialize μ_k
  4.   update $t = t + 1$
  5.   calculate membership matrix using Eq. (5)
  6.   calculate cluster centroid using Eq. (6)
  7.   if a stopping, i.e., $t > T$ or $\|R^t - R^{t-1}\|_F < \varepsilon$, is fulfilled then stop,
       else go back to step 4
  8.   calculate cluster centroid from original data with multiplying R and
       original data X

3   Results and Discussion

In this section, we are going to analyze and compare the accuracies of the RFCM and
KRFCM with the Eigenspace-based Fuzzy C-Means (EFCM) [11] and Kernelized
Eigenspace-based Fuzzy C-Means (KEFCM) [13] method. The comparison uses a
measurement unit called *topic interpretability*.

3.1   Topic Interpretability

Topic Interpretability is a quantitative method to seek the interpretability of a topic by
calculating the coherence scores of words that construct the topic. One of the standard
formulas which used to estimate coherence scores is PMI. Suppose a topic t consists of
an n-word that is $\{w_1, w_2, \ldots, w_n\}$, the PMI of the topic t is

$$PMI(t) = \sum_{j=2}^{n} \sum_{i=1}^{j-1} \log\left(\frac{P(w_j, w_i)}{P(w_i)P(w_j)}\right) \tag{8}$$

where $P(w_j, w_i)$ is the probability of the word w_i appears together with the word w_j on
the corpus, $P(w_i)$ is the probability of the word w_i appears in the corpus, and $P(w_j)$ is
the probability of the word w_j appears in the corpus. Corpus is a database consisting of
some text-based documents which used for a reference to calculate PMI. In this
experiment, we use a corpus consisting of 3.2 million Indonesian Wikipedia
documents.

In this simulation, we use the dataset from 14 Indonesian national news account on twitter with the majority are the Indonesian language. The dataset is gathered by using a streaming method from March 4th, 2019 until March 9th, 2019. The streaming method itself is done by taking the tweets after the tweets are being made. The benefit of the streaming method is that the data gathered are live tweets. The datasets consist of 5425 tweets and 11056 words. Table 1 below provides some additional information about the dataset used in this paper.

Table 1. Additional Information about the dataset

Number of tweets	5425
Number of words	11056
Number of national news account	14
Streaming start (DD/MM/YYYY HH:MM:SS)	04/03/2019 15:07:28 GMT + 07
Streaming end (DD/MM/YYYY HH:MM:SS)	09/03/2019 06:36:14 GMT + 07
Streaming duration (hours)	111.5

The datasets have been in the form of a list of sentences. So, the process that needs to be done is only to form the word-document matrix X and do the weighting process. In this simulation, the weighting process is performed using the term frequency-inverse document frequency (TF-IDF) [16].

To convert text into vector representations, we do some pre-processing. Firstly, we delete link on every tweet; convert some words with the hashtag to the separate words; convert all words into lowercase; replace two or more repeating letters with only two occurrences; delete all the numbers on tweets; delete every username on tweets; delete the words with two or fewer characters. For the final step, we used the term frequency-inverse document frequency for weighting the dataset. Besides using scikit-learn packages for preprocessing and tokenizing steps, we also used the natural language toolkit from nltk [17].

In this simulation, we also inspect two other methods for topic detection; they are EFCM [11] and KEFCM [13]. For EFCM, KEFCM, RFCM, and KRFCM parameters, we arrange the fuzzification constant $w = 1.5$, the maximum number of iteration $T = 1000$, the threshold $\varepsilon = 0.005$ and we are reducing its dimension into 5 using Truncated SVD for EFCM and KEFCM and Gaussian Random Projection for RFCM and KRFCM. As for KEFCM and KRFCM, we use $\gamma = 0.001$. All of these methods are repeated five times.

These methods produce topics consisting of 10 words, in each topic forming one vector. And after the results are collected, then the words in each topic are going to calculated by PMI in Eq. 8 to get coherence score.

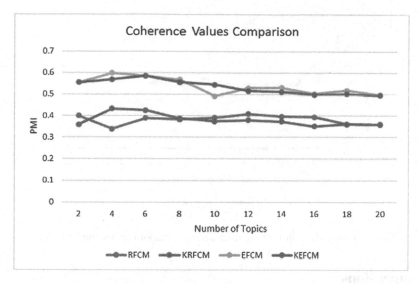

Fig. 1. Comparison in PMI score between RFCM, KRFCM, EFCM, and KEFCM

Figure 1 shows the comparison of Coherence scores for the number of topic $c \in \{2, 4, \ldots, 18, 20\}$. Based on Fig. 1, we can see that RFCM and EFCM have smaller PMI values than EFCM and KEFCM methods. Table 2 below gives a summary of the PMI scores for each method. EFCM method achieves the highest optimal PMI scores of the other three methods.

Table 2. The comparison of the optimal topic interpretability (PMI) scores

RFCM	KRFCM	EFCM	KEFCM
0.4087	0.4331	0.5471	0.5468

Figure 2 below shows that the random projection outperformed the SVD in terms of running time.

Analytically, the running time of the random projection can be denoted as $O(c \times k \times N)$ where c states the number of entities not empty per column, k denotes the dimensions after being reduced, and N denotes the amount of data [18]. While the running time of the SVD can be denoted as $O(n_{\text{max}}^2 \times k)$ with $n_{max} = max(n_{document}, n_{feature})$ where the $n_{document}$ states the number of documents and $n_{feature}$ state the number of different words and k denotes the dimensions after being reduced[1].

---

[1] https://scikit-learn.org/stable/modules/decomposition.html.

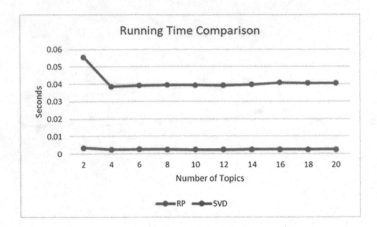

Fig. 2. Comparison in running time between random projection and SVD

4   Conclusions

In this paper, we examine randomspace-based fuzzy c-means and kernelized randomspace-based fuzzy c-means for topic detection on twitter. We use inter-pretability of the extracted topics in the form of coherence scores to compare RFCM and KRFCM methods with EFCM and KEFCM methods. Our simulations show that RFCM gives best results for 12 topics, and KRFCM gives best results for 4 topics. However, Both the RFCM and KRFCM learning methods cannot outperform the average PMI values produced by the EFCM or KEFCM learning methods, but RFCM and KRFCM excel at the time needed to reduce the dimensions of the dataset.

Acknowledgment. This work was supported by Universitas Indonesia under PIT 9 2019 grant. Any opinions, findings, and conclusions or recommendations are the authors' and do not nec-essarily reflect those of the sponsor.

References

1. Xie, W., Zhu, F., Jiang, J., Lim, E.-P., Wang, K.: Topic sketch: real-time bursty topic detection from Twitter. IEEE Trans. Knowl. Data Eng. **28**(8), 2216–2229 (2016)
2. Craig, T., Ludloff, E.M.: Privacy and Big Data. O'Reilly Media Inc., Sebastopol (2011)
3. Aiello, L.M., et al.: Sensing trending topics in Twitter. IEEE Trans. Multimedia **15**(6), 1268–1282 (2013). https://doi.org/10.1109/TMM.2013.2265080
4. Blei, D.M.: Probabilistic topic models. Commun. ACM **55**(4), 77–84 (2012)
5. Petkos, G., Papadopoulos, S., Kompatsiaris, Y.: Two-level message clustering for topic detection in Twitter. In: Proceedings of the SNOW 2014 Data Challenge, Seoul, Korea, 8 April 2014 (2014)

6. Nur'aini, K., Najahaty, I., Hidayati, L., Murfi, H., Nurrohmah, S.: Combination of singular value decomposition and k-means clustering method for topic detection on Twitter. In: Proceedings of International Conference on Advanced Computer Science and Information System, Depok, Indonesia, 10–11 October 2015 (2015)
7. Fitriyani, S.R., Murfi, H.: The k-means with mini batch algorithm for topics detection on online news. In: Proceedings of the 4th International Conference on Information and Communication Technology, Bandung, Indonesia, 25–27 May 2016 (2016)
8. Bezdek, J.C.: Pattern Recognition with Fuzzy Objective Function Algorithms. Platinum Press, New York (1981)
9. Daniel, G., Witold, P.: Kernel-based fuzzy clustering and fuzzy clustering: a comparative experimental study. Fuzzy Sets Syst. **161**(3), 522–543 (2010). https://doi.org/10.1016/j.fss.2009.10.021
10. Winkler, R., Klawonn, F., Kruse, R.: Fuzzy c means in high dimensional spaces. Int. J. Fuzzy Syst. Appl. **1**, 1–16 (2011)
11. Muliawati, T., Murfi, H.: Eigenspace-based fuzzy c-means for sensing trending topics in Twitter. In: AIP Conference Proceedings, vol. 1862, no. 1, July 2017. http://doi.org/10.1063/1.4991244
12. Murfi, H.: The accuracy of fuzzy c-means in lower-dimensional space for topic detection. In: Qiu, M. (ed.) SmartCom 2018. LNCS, vol. 11344, pp. 321–334. Springer, Cham (2018). https://doi.org/10.1007/978-3-030-05755-8_32
13. Prakoso, Y., Murfi, H., Wibowo, A.: Kernelized eigenspace based fuzzy C means for sensing trending topics on Twitter. In: Proceedings of the International Conference on Data Science and Information Technology, Singapore (2018)
14. Vu, K.K.: Random projection for high-dimensional optimization. Optimization and Control. Université Paris-Saclay. English (2016)
15. Johnson, W.B., Lindenstrauss, J.: Extensions of Lipshitz mapping into Hilbert space. In: Conference in Modern Analysis and Probability. Contemporary Mathematics, vol. 26, pp. 189–206. American Mathematical Society (1984)
16. Manning, C.D., Schuetze, H., Raghavan, P.: Introduction to Information Retrieval. Cambridge University Press, Cambridge (2008)
17. Loper, E., Bird, S.: NLTK: the natural language toolkit. In: Proceedings of the COLING/ACL Interact. Present. Sess, pp. 69–72 (2006)
18. Bingham, E., Mannila, H.: Random projection in dimensionality reduction. In: Proceeding of the Seventh ACM SIGKDD International Conference on Knowledge Discovery and Data Mining. ACM, New York (2001)

Image Stitching Based on Discrete Wavelet Transform and Slope Fusion

Daochen Weng, Qianying Zheng$^{(\boxtimes)}$, and Bingkun Yang

College of Physics and Information Engineering, Fuzhou University,
Fuzhou 350116, China
zhengqy@vip.sina.com

Abstract. The fusion algorithm of traditional image stitching does not fully consider the differences of the clarity of the two images, and the conventional Discrete Wavelet Transform algorithm would blur the image when applied to image stitching. Owing to these, an improved method based on Discrete Wavelet Transform and Slope Fusion is proposed. The proposed algorithm firstly performs Haar wavelet transform on the image to be fused to obtain a low-frequency component and multiple high-frequency components. Subsequently, the Slope Fusion method is used for the obtained low-frequency component and the sub-regional Slope Fusion method is used for the high-frequency components. Finally, the fused image is obtained by using the Inverse Discrete Wavelet Transform for the new low-frequency component and high-frequency components. The proposed algorithm can retain the information of direction and detail while taking full account of differences in image sharpness, all of those benefits help improve the quality of the fused image effectively. The experimental results show that the proposed algorithm can make the fused image clearer and objectively enhance multiple fusion indicators of the fused image.

Keywords: Image stitching · Discrete Wavelet Transform · Slope Fusion · Inverse Discrete Wavelet Transform · Fusion indicators

1 Introduction

In the past, in order to obtain images with wide field of view, expensive cameras with high-resolution are usually required, which posed a huge challenge to the hardware. The proposed image stitching technology effectively solves this problem to a certain extent. Image stitching [1] is a technology which combines several images with overlapping areas into a seamless panorama or a high-resolution image. It is widely used in computer vision [2], medical image analysis [3], remote sensing image [4] and virtual reality [5].

Image stitching is mainly divided into three parts: image preprocessing, image registration and image fusion. The purpose of image preprocessing is to eliminate the interference as much as possible before stitching. Common image preprocessing methods include image filtering, projection transformation [6,7], etc. Image

© Springer Nature Switzerland AG 2019
R. Chamchong and K. W. Wong (Eds.): MIWAI 2019, LNAI 11909, pp. 144–155, 2019.
https://doi.org/10.1007/978-3-030-33709-4_13

registration is the key to mapping two images to the same coordinate. Among all the current algorithms, the SURF [8] (Speeded-Up Robust Features) algorithm performs the best in image registration. The quality of image fusion directly influences the final stitching effect. In all image fusion algorithms, the pixel-level fusion [9] works best at present.

For two images to be stitched, one is generally used as a reference image and the other as a target image. The pixel-level fusion is to mix pixel points at the same target position in the reference image and the target image by certain rules. Conventional pixel-level fusion algorithms include Comparison Fusion (CF), Average Fusion (AF) and Gradual Fusion (GF). Among them, the CF and the AF will lead to noticeable stitching seams in stitching images, and the GF algorithm can solve this problem with effect. Since we usually use the reference image to calculate the homography matrix of the target image and then perform the affine transformation on the target image, which will lead to the loss of accuracy of the target image. However, the traditional GF algorithm whose weight of fusion point is linearly related to the distance between the two boundaries of fusion region, since this does not fully consider the difference in sharpness between the two images, which will lead to insufficient details of the fused image. For more consideration of the details of the reference image, an improved Slope Fusion (SF) algorithm is proposed in this paper. The proposed algorithm takes a larger weight on the reference image, which can effectively improve the details of the fused image and make the stitching image clearer.

The pixel-level fusion commonly uses the Discrete Wavelet Transform [10,11] (DWT) as a tool for multi-scale fusion, which can preserve the information of direction and detail of images. After the DWT, a low-frequency component representing the overall contour and luminance information of the image and multiple high-frequency components representing the details of the image are obtained. The low-frequency component of the DWT usually use the CF method or the AF method, which also lead to obvious stitching seams, and the high-frequency components of the DWT usually use the rules of Regional Characteristic Measurement [12,13], which will cause the image blurred.

In this work, an improved algorithm based on the DWT and the SF (DWT-SF) is proposed. The algorithm performs SF on the low-frequency component and the high-frequency components after the DWT, which effectively preserves more image details and makes the merged image clearer. The structure of this paper is organized as follows. The principle of traditional image fusion algorithm is briefly introduced in Sect. 2. In Sect. 3, an improved DWT-SF algorithm which combines the DWT and SF method is introduced. Experiment results are present in Sect. 4. Finally, conclusions are given in Sect. 5.

2   Traditional Image Fusion Algorithm

Image fusion refers to mix image data about the same target position from two images into one image, which makes the fused image rich in more useful information. The equation of image fusion is as follows:

$$I(x,y) = \omega_1 I_1(x,y) + \omega_2 I_2(x,y) \tag{1}$$

Where $I_1(x,y)$ and $I_2(x,y)$ represent the pixel value at the same target position in the reference image and the target image respectively, ω_1 and ω_2 represent the weight coefficients of the pixel value at that location of the reference image and the target image respectively, and $\omega_1 + \omega_2 = 1$, $0 \le \omega_1 \le 1$, $0 \le \omega_2 \le 1$.

2.1 Gradual Fusion

Assume that the width of the fusion region is d, d_1 is the distance between $I_1(x,y)$ and the left boundary of the fusion region. The weight coefficient equation for the traditional Gradual Fusion is as follows:

$$\omega_1 = \begin{cases} 1 & (x,y) \in I_1 \\ d_1/d & (x,y) \in I_1 \cap I_2 \\ 0 & (x,y) \in I_2 \end{cases} ; \omega_2 = 1 - \omega_1 \tag{2}$$

The variation of weights of Gradual Fusion is shown in Fig. 1

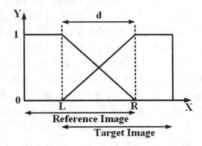

Fig. 1. The variation of weights of Gradual Fusion.

2.2 Wavelet Fusion

The wavelet fusion algorithm is flexible and adaptive, it can perform multi-scale decomposition on images for post-processing. The wavelet fusion can also improve image details. It can ensure faster fusion speed and better fusion effect at the same time.

The selection of fusion rules is the most important part of wavelet fusion algorithm based on the DWT. An excellent fusion rule preserves more image details and makes the image clearer. The basic framework of wavelet fusion is shown in Fig. 2.

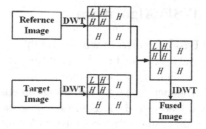

Fig. 2. The basic framework of wavelet fusion.

Haar Wavelet Transform. The simplest DWT is the Haar wavelet transform [14]. Haar wavelet transform guarantees processing speed without loss of image details. Taking one row of pixels in the source image as an example, the principle of Haar wavelet transform is shown in Fig. 3.

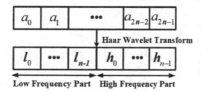

Fig. 3. The principle of Haar wavelet transform.

Suppose that one-row pixel data of the source image is $a_0 \sim a_{2n-1}$, the low-frequency part after wavelet transform is $l_0 \sim l_{n-1}$, and the high-frequency part is $h_0 \sim h_{n-1}$. The calculation criterion is as follows:

$$\begin{cases} l_i = (a_{2i} + a_{2i+1})/2 \\ h_i = (a_{2i} - a_{2i+1})/2 \end{cases} \quad i = 0, 1, ..., n-1 \tag{3}$$

According to the above calculation rule, each row of the source image is subjected to an operation to obtain a one-dimensional Haar wavelet transform result. Since the image is two-dimensional, it is necessary to perform another Haar wavelet transform in the direction of the column after the row transformation to get a two-dimensional Haar wavelet transform result.

Inverse Haar Wavelet Transform. After fusion, the image needs to be restored using the inverse Haar wavelet transform. The inverse Haar wavelet transform is the process of solving linear equations. The equation for the inverse Haar wavelet transform which is easily derived by Eq. 3 is:

$$\begin{cases} a_{2i} = l_i + h_i \\ a_{2i+1} = l_i - h_i \end{cases} \quad i = 0, 1, ..., n-1 \tag{4}$$

3   Improved DWT-SF Algorithm

The traditional Gradual Fusion algorithm whose weight of fusion point is linearly related to the distance between the two boundaries of the fusion region. In other words, the weight coefficients of the reference image and the target image are actually equivalent. In general, the target image needs to undergo an affine transformation, which will lead to a decrease in its sharpness, while the Gradual Fusion does not consider this aspect. In order to make the fused image as clear as possible, it should focus more on the weight of the reference image. Based on this, a Slope Fusion algorithm is proposed to improve the weight selection part of the Gradual Fusion in this paper. Then, combining the Slope Fusion algorithm with the DWT algorithm, an improved DWT-SF algorithm is proposed.

3.1   Improved Slope Fusion Algorithm

Assuming that d_1 is the distance between $I_1(x, y)$ and the left boundary of the fusion region, and the width of the fusion region is d, The weight coefficient equation for the Slope Fusion is as follows:

$$\omega_1 = \begin{cases} 1 & (x, y) \in I_1 \\ 1 - d_1{}^3/d^3 & (x, y) \in I_1 \cap I_2 \\ 0 & (x, y) \in I_2 \end{cases} ; \omega_2 = 1 - \omega_1 \tag{5}$$

The variation of weights of Slope Fusion is shown in Fig. 4.

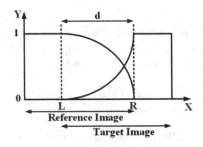

Fig. 4. The variation of weights of Slope Fusion.

3.2   Low-Frequency Component Using Slope Fusion Algorithm

After N-times decomposition, there is only one low-frequency component representing the overall luminance information of the image. Most algorithms use Comparison Fusion method for low-frequency component, which causes obvious bright and dark changes in the fused image. There are also some algorithms that use the Average Fusion method for low-frequency component, which will result in noticeable stitching seams.

Aiming at the problem of the unsatisfactory fusion effect when using the Comparison Fusion method or the Average Fusion method for low-frequency component, in this paper, the Gradual Fusion algorithm and the Slope Fusion algorithm are applied to the fusion of low-frequency component. Comparing the fusion results with Comparison Fusion and Average Fusion, the visual effect is obviously better. However, as mentioned above, the Gradual Fusion algorithm does not adequately consider the difference in sharpness between the two images. Based on this consideration, the Slope Fusion algorithm is chosen as the fusion rule for the low-frequency component.

3.3   High-Frequency Components Using Sub-regional Slope Fusion Algorithm

After N-times decomposition, there are $3N$ high-frequency components, which represent the details of the image. At present, the fusion of high-frequency components mostly adopts the rule of Regional Characteristic Measurement. When this method is applied to color image fusion, the details of the image will be quite ambiguous. This is because the weight of the high-frequency coefficient is only related to the pixel itself and its surrounding pixels, which is independent of the spatial position of the pixel in the fusion region. Its essence is a disguised Comparison Fusion.

In this paper, the sub-regional Slope Fusion algorithm is used for the high-frequency components, which makes the fused image clear. This method divides the high-frequency components into multiple independent regions, then using Slope Fusion algorithm for each region. This method takes into account the spatial position of each high-frequency region at each scale and assigns weights accordingly. If the entire high-frequency area is treated as a whole and only one Slope Fusion algorithm is performed, the effect will become extremely poor. The specific rules are as follows:

4   Experiment Results

Before image stitching, the reference image and the target image are first transformed by cylindrical projection transformation. Then, using the SURF algorithm to extract feature points from the projected transformed image. After that, the KNN (k-Nearest Neighbors) common matching [15] algorithm is used to perform rough matching on feature points and then refined by RANSAC [16] (Random Sample Consensus) algorithm. Subsequently, the homography matrix is calculated by using the purified feature points. Next, using the obtained homography matrix to perform the affine transformation on the target image. Finally, find the best suture in the overlapping area of the image after registration and extract the fusion area near the optimal suture. After all the steps are completed, the fusion algorithms are used for fusion in the fusion region, then the algorithms are compared and analyzed.

In order to make the experimental results clearer, this experiment uses two images with a resolution of 3284 × 1840 for stitching. Figure 5(a) is the reference image, Fig. 5(b) is the target image, and Fig. 5(c) shows the result of stitching using the improved DWT-SF algorithm.

(a)   (b)

(c)

Fig. 5. Experimental images and stitching result: (a) reference image, (b) target image, and (c) stitching result.

4.1 Fusion Indicators

In order to verify the effectiveness of the proposed algorithm, three objective fusion indicators are used in the experiment:

(1) Information Entropy (IE): Information entropy reflects how much information the image contains. The greater the information entropy, the richer the amount of information contained in the image. Information entropy is defined as:

$$E = -\sum_{i=0}^{L-1} p_i \log_2 p_i \tag{6}$$

Where L represents the total gray level of the image, and p_i represents the ratio of the number of pixels to the total number of pixels of the image whose gradation value is i.

(2) Average Gradient (AG): The average gradient reflects the ability to express minute detail contrast and texture change features while reflecting the sharpness of the image. The larger the average gradient, the better the visualization and resolution of the image. The average gradient is defined as:

$$AG = \frac{1}{(M-1)(N-1)} \sum_{x=1}^{M} \sum_{y=1}^{N} \sqrt{\frac{1}{2}\left[(\frac{\partial I(x,y)}{\partial x})^2 + (\frac{\partial I(x,y)}{\partial y})^2\right]} \tag{7}$$

Where M and N are the width and height of the image, respectively. $I(x, y)$ is the gray value of the pixel coordinated with (x, y).

(3) Spatial Frequency (SF): The spatial frequency reflects the overall activity of the image space domain. The larger the spatial frequency of the image, the better the fused image. The spatial frequency is defined as:

$$SF = \sqrt{RF^2 + CF^2} \tag{8}$$

Where RF is the line frequency of the image and CF is the column frequency of the image, which are defined as:

$$RF = \sqrt{\frac{1}{M \times N} \sum_{x=1}^{M} \sum_{y=2}^{N} [I(x, y) - I(x, y - 1)]^2} \tag{9}$$

$$CF = \sqrt{\frac{1}{M \times N} \sum_{x-2}^{M} \sum_{y-1}^{N} [I(x, y) - I(x - 1, y)]^2} \tag{10}$$

Where M and N are the width and height of the image, respectively. $I(x, y)$ is the gray value of the pixel coordinated with (x, y).

4.2 Comparison of Wavelet Fusion Rules

Comparison of Fusion Rules for Low-Frequency Component. For the two images in Fig. 5(a) and (b), the high-frequency components are fused by the sub-regional Slope Fusion method, and the low-frequency component is fused by using the Comparison Fusion, the Average Fusion, the Gradual Fusion and the Slope Fusion, Separately. Then the local regions are extracted for comparing, as shown in Fig. 6.

The experimental results show that using the Comparison Fusion method and the Average Fusion method for the low-frequency component will result in obvious stitching seams, and the details are relatively vague. The Gradual Fusion and the Slope Fusion effectively remove the stitching seams, and the details are clearer. Therefore, it is better to use the Gradual Fusion method and the Slope Fusion method for the low-frequency component.

Comparison of Fusion Rules for High-Frequency Components. For the two images in Fig. 5(a) and (b), the low-frequency component is fused by the Slope Fusion method, and the high-frequency components are subjected to the Regional Characteristic Measurement method, the sub-regional Gradual Fusion, and the sub-regional Slope Fusion, Separately. Then the local regions are extracted for comparing, as shown in Fig. 7.

The experimental results show that the fusion rules based on Regional Characteristic Measurement can make the fused image quite fuzzy. The images obtained by the sub-regional Gradual Fusion method and the sub-regional Slope Fusion method are obviously clearer. Therefore, it is better to use the sub-regional Gradual Fusion method and the sub-regional Slope Fusion method for the high-frequency components.

4.3   Comparison of Fusion Indicators

When the Gradual Fusion algorithm and the Slope Fusion algorithm are used for the low-frequency component and the high-frequency components, it is difficult to judge the effect of the fusion intuitively. Therefore, objective fusion indicators are necessary to be used for further judgment. In order to compare the advantages and disadvantages of direct fusion and wavelet fusion, this experiment uses the same fusion rules for low-frequency component of wavelet and high-frequency components, in other words, both of them using the Gradual Fusion or the Slope Fusion.

In experiment, four methods are used for fusion, which are the DWT-SF method, the DWT-GF (the combination of DWT and Gradual Fusion) method, the SF method and the GF method separately.

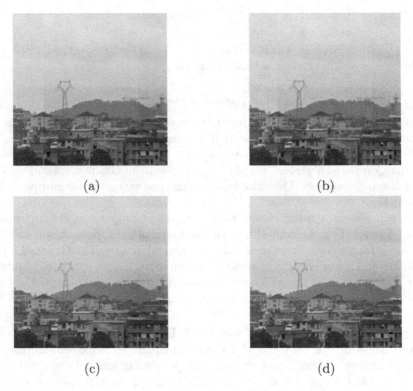

Fig. 6. Comparison of Fusion Rules for Low-Frequency Component: (a) Comparison Fusion, (b) Average Fusion, (c) Gradual Fusion, and (d) Slope Fusion.

For the two images in Fig. 5(a) and (b), the fusion regions are fused using the four methods mentioned above, and the fusion indicators are calculated. The results are shown in Table 1.

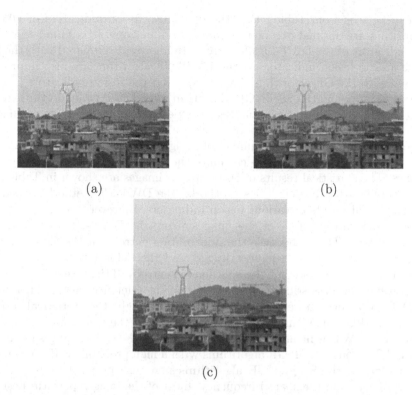

(a)   (b)

(c)

Fig. 7. Comparison of Fusion Rules for High-Frequency Components: (a) Regional Characteristic Measurement Fusion, (b) sub-regional Gradual Fusion, and (c) sub-regional Slope Fusion.

Table 1. The fusion indicators of stitching results of Fig. 5(a) and (b).

Solution	Information entropy	Average gradient	Spatial frequency
DWT-SF	7.446993	5.201268	12.423985
DWT-GF	7.457325	4.704870	10.978650
SF	7.439032	5.172541	12.379979
GF	7.452849	4.690785	10.945019

Table 2. The statistical results of 50 groups of images.

Solution	Information entropy	Average gradient	Spatial frequency
DWT-SF	88%	94%	90%
DWT-GF	12%	2%	4%
SF	0%	2%	6%
GF	0%	2%	0%

It can be seen from Table 1 that the other two fusion indicators of DWT-SF algorithm are optimal except the information entropy. It is found that the fusion indicators of the DWT-SF algorithm are improved compared with the SF algorithm. The fusion indicators of the DWT-GF algorithm are also improved compared with the GF algorithm. This shows that the fusion indicators can be improved by applying the SF or GF algorithm after DWT to a certain extent. Due to the low credibility of just one set of experimental data, 50 sets of images are stitched using these four methods and then calculate their fusion indicators for comparison. First, count the number of images with the best fusion indicator among the four methods, and then calculate the proportion of them in 50 groups of images. The statistical results of 50 groups of images are shown in Table 2.

Compared with the other three methods, the DWT-SF method is optimal in the statistical results of various fusion indicators. However, for the indicator of information entropy, 12% of the images using the DWT-GF method have reached the best. This is because the information entropy of the target image is larger than that of the reference image and the deformation of the affine transformation is small, so that the information entropy of the target image after affine transformation is still larger than that of the reference image. Therefore, the DWT-GF algorithm performs the best. Interestingly, the statistical results for the SF and GF algorithms are 0%, this is because the fusion indicators will improve after DWT as mentioned above. This makes the statistical results focus on the DWT-SF and DWT-GF algorithms with a high probability. If the sample size is increased, the SF and GF algorithms may also perform best. For the average gradient and the spatial frequency, most of the images perform best on the DWT-SF algorithm. In summary, the DWT-SF algorithm proposed in this paper is more advantageous.

5   Conclusion

In this work, we first analyze the advantages and disadvantages of the traditional GF algorithm and the DWT algorithm. In view of the shortcomings of the GF algorithm, we propose an improved SF algorithm, and then combine it with the DWT to propose the DWT-SF algorithm. The fusion algorithm proposed in this paper is compared with the traditional fusion algorithm. It is found that the proposed algorithm can intuitively keep the merged image clear. Then randomly select 50 groups of images for stitching, and compare the four algorithms with good visual fusion effects. The experimental results show that the proposed algorithm can also perform well on objective fusion indicators.

Acknowledgements. The authors would like to acknowledge the supports by the National Natural Science Foundation of China (Grant No. 61471124), Key Industrial Guidance Projects of Fujian Science and Technology Department (Grant No. 2016H0016 and 2015H0021).

References

1. Ghosh, D., Kaabouch, N.: A survey on image mosaicing techniques. J. Vis. Commun. Image Represent. **34**, 1–11 (2016)
2. Ha, Y.J., Kang, H.D.: Evaluation of feature based image stitching algorithm using OpenCV. In: 10th International Conference on Human System Interactions (HSI) (2017)
3. Chen, X., Liu, H., Zhou, M., et al.: Medical image mosaic based on low-overlapping regions. In: International Congress on Image and Signal Processing (2018)
4. Zhang, W., Li, X., Yu, J., et al.: Remote sensing image mosaic technology based on SURF algorithm in agriculture. EURASIP J. Image Video Process. **2018**(1), 1–9 (2018)
5. Ling, Y., Yong, C., Yun, C.: The key technology of virtual reality system based on panoramic view. Appl. Mech. Mater. **130–134**, 3123–3127 (2011)
6. Chen, K., Wang, M.: Image stitching algorithm research based on OpenCV. In: Control Conference (2014)
7. Lin, M., Xu, G., Ren, X., et al.: Cylindrical panoramic image stitching method based on multi-cameras. In: IEEE International Conference on Cyber Technology in Automation (2015)
8. Bay, H., Ess, A., Tuytelaars, T., et al.: Speeded-Up Robust Features (SURF). Comput. Vis. Image Underst. **110**(3), 346–359 (2008)
9. Li, S., Kang, X., Fang, L., et al.: Pixel-level image fusion: a survey of the state of the art. Inf. Fusion. **33**, 100–112 (2017)
10. Li, S., Kwok, J.T., Wang, Y.: Using the discrete wavelet frame transform to merge Landsat TM and SPOT panchromatic images. Inf. Fusion **3**(1), 17–23 (2002)
11. Pajares, G., Cruz, J.M.D.L.: A wavelet-based image fusion tutorial. Pattern Recognit. **37**(9), 1855–1872 (2004)
12. Burt, P.J., Adelson, E.H.: Merging images through pattern decomposition. In: Applications of Digital Image Processing VIII (1985)
13. Zhang, B.: Study on image fusion based on different fusion rules of wavelet transform. In: Proceedings of the 3rd International Conference on Advanced Computer Theory and Engineering (ICACTE) (2010)
14. Daza, R.J.M., Upegui, E.: Image fusion using the wavelet TRW and Haar transforms: enhancement of spatial resolution for the Ikonos images from Ortophotos. In: 7th IEEE International Conference on Software Engineering and Service Science (ICSESS) (2016)
15. Qu, Z., Bu, W., Liu, L., et al.: The algorithm of seamless image mosaic based on A-KAZE features extraction and reducing the inclination of image. IEEJ Trans. Electr. Electron. Eng. **13**(1), 134–146 (2018)
16. Fischler, M.A., Bolles, R.C.: Random sample consensus: a paradigm for model fitting with applications to image analysis and automated cartography. Commun. ACM **24**(6), 381–398 (1981)

Dynamic Hand Gesture Recognition from Multi-modal Streams Using Deep Neural Network

Thanh-Hai Tran[1]([envelope]) [iD], Hoang-Nhat Tran[1] [iD], and Huong-Giang Doan[2] [iD]

[1] International Research Institute MICA, Hanoi University of Science and Technology, Hanoi, Vietnam
thanh-hai.tran@mica.edu.vn, hnhat.tran@gmail.com
[2] Electric Power University, Hanoi, Vietnam
giangdth@epu.edu.vn

Abstract. Hand gesture is an efficient mean of human computer interaction. However, hand gesture recognition faces many challenges such as low hand resolution, phase variation and viewpoint. As a result, deployment of hand gesture in a practical application of human machine interaction is still very limited. This work aims to increase performance of hand gestures recognition by using multi-modal streams. We propose a method that combines depth, RGB and optical flow in a unified recognition framework. Each stream will go first into a feature extraction component, which is a deep learning model. We then investigate different fusion techniques to combine features from multi-modal streams for final classification. The proposed method is validated on a dataset of twelve gestures collected by ourselves from five different viewpoints. Experimental results show that accuracy of the proposed method using multi-modal streams outperforms ones that use a single stream, particularly for difficult viewpoints.

Keywords: Hand gesture recognition · Multi-modal fusion · Deep learning

1 Introduction

Recently, there was a significant growth of novel devices and techniques for human machine interaction. These allow new modalities of communication (e.g. gesture, speech) which are becoming more and more intuitive and easy to use. Among those modalities, hand gestures have gained much of attention of researchers and developers in recent years. Although there exists a number of methods of hand gestures recognition, deploying such a method in a practical application remains a problem due to many challenges: low resolution of hands,

This material is based upon work supported by the Air Force Office of Scientific Research under award number FA2386-17-1-4056.

phase variation of gestures, cluttered background, viewpoint changes. In a context of controlling home appliances using hand gestures, a perfect system must allow the end-user to stand any where and toward any direction w.r.t. the sensor. In those cases, the problem of scale and viewpoint become more and more critical.

Depending on the characteristic of data streams, existing methods belong to three categories: RGB based methods, depth based methods and hybrid methods. RGB based methods have been studied long-times ago due to the popularity of conventional RGB camera [14]. In the last decade, depth sensors (Microsoft Kinect, Prime Sense) were introduced and widely used nowadays due to its low-cost. These sensors provide not only RGB but also depth data which is more robust to lighting condition and could be a complementary for video representation. Then a number of methods for hand gestures using depth have been proposed [9,15,16]. However, sole depth could be unreliable when the subject is out of measurement range or noisy. Mostly when hand is of low resolution, the use of sole depth may be unsuitable [4,5].

In this paper, we are concerned with visual interpretation of hand gestures, and we attempt to improve performance of current methods of human hand gestures recognition using multiple modalities. In the literature, most of works combine RGB with depth [11] or RGB with optical flow [2]. In [2], combining RGB with optical flow obtained better performance. In [8], the authors indicated that algorithms using depth outperformed RGB. Our questions are: *Does combination of all of these modalities (RGB, depth, optical flow) boost recognition outcome? Does depth still help increasing recognition performance when hand has low resolution?*

Our proposed method takes depth, RGB and optical flow as three main streams. The framework consists of three components. For feature extraction, we deploy a deep neural network C3D [18] which has been shown to be an efficient feature extractor of actions. Each data stream will be fed to a C3D for feature extraction. To combine multi-modal features, we investigate different fusion techniques including early fusion, late fusion and a multiple kernel learning (MKL) technique. Finally, Support Vector Machine (SVM) is utilized for gesture classification. To evaluate the proposed methods, we collect a dataset of twelve hand gestures of twelve people from five different viewpoints.

In summary, our main contribution is three-folds: a three-streams framework for gestures recognition beyond deep neural network; an investigation of different fusion techniques for human hand gesture recognition; a validation of the proposed method on a dataset of hand gestures and evaluation of the impact of viewpoint changes on recognition performance of the proposed method. The remaining of this paper is as follow. In Sect. 2, we present related works on human hand gesture recognition and fusion methods. In Sect. 3, we describe our proposed method. Dataset and experiments are presented in Sect. 4. Conclusion will be mentioned in Sect. 5.

2   Related Works

In recent years, there has been a number of works on hand gesture recognition [14]. Methods for hand gesture recognition could use single or multiple modalities and usually fall into two main categories: hand crafted feature based approaches [5,12,19] and deep learning based approaches [2,7,18]. We will review these methods in the following sub-sections.

Hand Crafted Features Based Approaches. The methods belonging to these approaches involve extracting manually designed features for video representation from depth or RGB streams. [20] utilizes depth sequences by integrating Edge Enhanced Depth Motion Map (E^2DMM) with Histogram of Gradient (HOG) descriptor. [10] extracts three-dimensional (3D) spatial trajectory feature. [5] combines manifold learning (ISOMAP) to extract spatial feature and trajectory representation (KLT) for temporal information from segmented hand sequence. [4] improves Kernel Descriptor (KDES) to present for hand gesture which is adaptive to different scales of hand region. [9] proposes system with two different Cell occupancy feature and Silhouette feature descriptor on depth images. [20] proposes a distance metric for matching fingers called Finger-Earth Mover's Distance (FEMD).

Deep Learning Based Approaches: Recently, deep learning has showed its out-performance compared to hand crafted features methods in many tasks. In [8], C3D, ResNet-10, ResNeXt-101 were experimented on Depth, RGB. In [6], the fast neural network based representative frames is firstly applied with 3D CNN to classify and estimate key frames. Then, the deep neural network SegNet (based VSUMM) is used to extract features. In [3], the author introduce a new CNN to recognize hand gesture where sequences of hand-skeletal joints' positions are processed by parallel convolutions.

Multi-modal Fusion: As multiple streams could give better representation of video, different methods using both RGB, Depth or Optical Flow have been proposed. In [13], the authors extracted different handcrafted features from RGB and Depth streams then combined these features in a kernelized SVM. In [21], the authors used a C3D neural network and a LSTM for feature extraction from each data stream (RGB and Depth). Then late fusion was adopted. In [11], a Focus of Attention net (FOANet) with late fusion was proposed. [2] combined RGB with Optical Flow by early and late fusion. In this paper, we would like to investigate which combination of RGB, Depth, Optical flow and which fusion techniques give the best recognition performance.

3   Proposed Method for Human Action Recognition from Multi-modal Data Streams

3.1   General Framework

We propose a framework that combines multi-modal data streams as illustrated in Fig. 1. It consists of several steps, depending on the fusion technique to

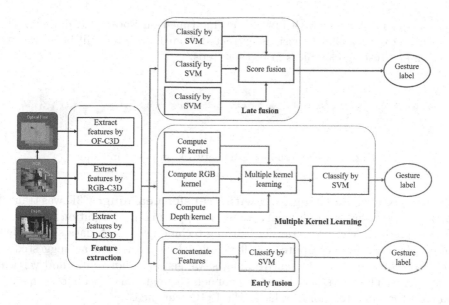

Fig. 1. Proposed framework for human activity recognition using multi-modal data streams. There are three fusion techniques to be investigated in our work. The figure is best seen in color. (Color figure online)

be used. Firstly, each data stream (RGB, Optical flow computed from RGB, Depth) goes into a feature extractor for extracting features. Next, these features extracted from each stream are combined using different fusion techniques. A classification technique SVM is applied for action label prediction. Components of our proposed framework will be described thoroughly in the following sections.

3.2 Feature Extraction from Multi-modal Data Streams

In this paper, we deploy a deep C3D model (3D deep convolution neural network), which was introduced in [18] and has shown to be very efficient for action recognition tasks. C3D takes input as an image sequence instead of a static image, computes the 3D convolution on each 3D cubes from video clip. By doing so, C3D captures both spatial and temporal characteristics of action at the same time. A LSTM (long short-term memory) module is not really necessary in C3D net. Authors in [2] utilized C3D for hand gesture recognition, moreover, only Optical flow and RGB are combined in late fusion schema while Depth data were not considered.

Summary of C3D Network: The C3D network contains 8 convolution, 5 maxpooling and 2 fully connected layers as showed in Fig. 2. The number of filters of convolution layers from Conv1 to Conv5 are 64, 128, 256, 512, 512 respectively. All 3D convolution kernels are of size $3 \times 3 \times 3$ with stride $1 \times 1 \times 1$. We utilize the network pre-trained with I380K and fine-tuned on Sports-1M dataset. The feature vector of 4,096 dimensions extracted from FC6 layer will be served for training and testing classifiers in further steps.

Fig. 2. Architecture of C3D network proposed in [18]

Feature Extraction Using C3D with Transfer Learning: C3D was trained on Sports-1M, a dataset of human actions. To apply in our recognition framework of hand gestures, we apply transfer learning technique on each data stream. For optical flow, we first compute this flow using Brox's method as suggested in [17]. From each pair of consecutive images, we generate horizontal and vertical components. Then we stack these two channels to create an optical flow sequence that has the same temporal order as the RGB sequence.

Regarding training details, we fine-tuned the original C3D models pre-trained with Sports-1M dataset. We started with learning rate of $1e{-}4$ for the first 500 iterations, then dropped it to $1e{-}5$ and continued for the next 300 iterations, both with RGB, optical flow and depth training data. Our training time is approximately three hours per modality on each view, with a NVIDIA GeForce GTX 1080 TI GPU. Finally, we obtained three models, namely RGB-C3D, OF-C3D and D-C3D for RGB, optical flow and depth stream respectively.

3.3   Modalities Fusion

In this work, we investigate different fusion techniques: early fusion, late fusion and multiple kernel learning. In the following sections, we briefly review each technique. We assume there is M modalities and L classes of gestures/actions. Each modality will go though a corresponding feature extractor and output a features vector $\mathcal{F}_i = [f_1^{(i)}, f_2^{(i)}, ..., f_{k_i}^{(i)}]$ with $i \in [1, M]$, k_i is the feature space dimension of the ith modality. In our work, k_i has the same value of 4,096.

Late Fusion: For late fusion, each feature vector from i^{th} modality will go through a SVM classifier $\mathcal{C}_i(\mathcal{F}_i)$ which outputs a score vector of L dimensions: $\mathcal{S}_i = [s_1^{(i)}, s_2^{(i)}, ..., s_L^{(i)}]$ with $i \in [1, M]$. Late fusion will take M score vectors $\{\mathcal{S}_i, i \in [1, M]\}$ and utilizes max or average operator for generating the final score vector $\mathcal{S}$ of L dimensions as formula (1):

$$\mathcal{S} = [op(s_1^{(1)}, ..., s_M^{(1)}), op(s_1^{(2)}, ..., s_M^{(2)}), ..., op(s_1^{(L)}, ..., s_M^{(L)})] = [s_1, s_2, ...s_L] \quad (1)$$

Where op is a certain operator (max or average). The final decision could be obtained from formula (2):

$$argmax_{i \in [1, L]} s_i \quad (2)$$

Early Fusion: For early fusion, each feature vector from the ith modality is normalized. Then the normalized features are concatenated into a final feature vector (F_f) as formula (3):

$$\mathcal{F}_f = [\mathcal{F}_1, \mathcal{F}_2, ..., \mathcal{F}_M] \tag{3}$$

The vector $\mathcal{F}_f$ will be inputted into a SVM classifier to predict the gesture/action label.

Multiple Kernel Learning: MKL algorithms were introduced for combining a set of base kernels representing different similarity measures of a sole data source or of different sources of data. In this paper, the second use case is considered, in which each feature vector from the ith modality is fed into a kernel function in order to compute the corresponding kernel matrix with formula (4):

$$K^{(i)} = similarity\left(f^{(i)}, l^{(i)}\right) \tag{4}$$

Where $l^{(i)}, f^{(i)}$ are two feature vectors in the ith modality. These kernel matrices $K^{(1)}, K^{(2)}, ..., K^{(M)}$ are then combined with formula (5):

$$\mathbf{K} = \theta_\eta(\{K^{(i)}\}_{i=1}^M | \eta) \tag{5}$$

Where θ is the functional form of the combination which can be linear sum $\sum_{i=1}^M \eta^{(i)} K^{(i)}$ or non-linear functions such as multiplication $\prod_{i=1}^M K^{(i)}{}^{\eta^{(i)}}$, and η denotes the combination coefficients, its values are bound to predefined rules or optimized by the learning process of MKL. In this work, a binary margin maximization MKL algorithm called EasyMKL [1] is chosen that uses convex summation function where all coefficients $\eta^{(i)}$ are restricted to be non-negative and sum to 1. A final kernel machine classifier will decide the label based on the combined kernel matrix $\mathbf{K}$. Our multi-class problem is mapped to multiple binary ones as we deploy one-vs.-rest strategy. Specifically, given L classes, L MKL modules are trained. With the set of L coefficients obtained, L output kernels are computed for training of L corresponding SVM classifiers.

4 Experiments

4.1 Dataset

In the context of using hand gestures for human machine interaction, we design a dataset which consists of twelve dynamic hand gestures $(G_1, G_2, ..., G_{12})$ corresponding to commands for controlling electronic home appliances (turn on/off, increase/decrease (speed/volume), swing, etc.). Each gesture is performed by moving hand following a pre-defined trajectory and changing hand shape in a natural manner. For each gesture, the hand starts from one position with closed posture, it opens and reshapes gradually during the first half cycle of movement, then recovers gradually to stop at its initial position and state. Figure 4

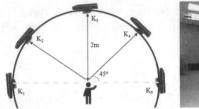

Fig. 3. Setup of Kinects for data acquisition. Five Kinects evenly spacing on half a circle are mounted on the shelf at height of 1.5 m (left). The subject stands right in front of Kinect 3 at distance of 2 m. On the right is the side view perspective given by Kinect 4. At the beginning of each gesture, the subject puts the hand at a position, performs the gesture then finishes at the same position at the beginning. We define a local coordinate system attached to the starting position of the hand for better illustration of the gestures in Fig. 4.

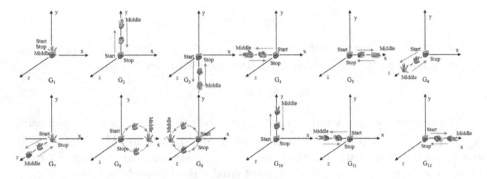

Fig. 4. Twelve defined dynamic hand gestures

illustrates the motions of hand and changes of postures during execution of twelve gestures. In this dataset, some pairs of gestures, for example (G_2, G_{10}), (G_4, G_{11}) and (G_5, G_{12}), have similar trajectory. The only difference is the hand shape during gesture implementation (e.g. Fig. 5). Even with human eyes, it is hard to distinguish these gestures at some viewpoints.

Five Kinect sensors K_1, K_2, K_3, K_4, K_5 are setup at five distinct positions, evenly spacing on half a circle whose center is the subject position and radius is the distance from the subject to the Kinects (in our setup is 2 m) (Fig. 3). All five Kinect sensors are synchronized to sample RGB and depth data at uniform rate. This setup allows us to capture a multi-view and multi-modal dataset. Twelve participants (8 males and 4 females) volunteered to perform the gestures. Each subject performs one gesture three to five times. We synchronously collected 570 gesture videos in total from each view. Compared to the dataset presented in [2], the number of gesture classes in this dataset is two times bigger and there exists firm similarity between some gestures.

This dataset will be publicly available soon. The environment setup of the dataset is illustrated in Fig. 3, the x-axis horizontally points from the right to the

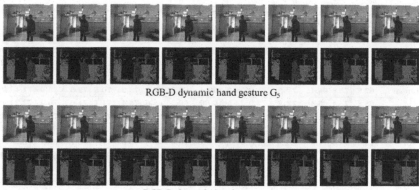

RGB-D dynamic hand gesture G_5

RGB-D dynamic hand gesture G_{12}

Fig. 5. Example of two similar gestures: gesture G_5 (two first rows show the RGB and depth key-frames) and gesture G_{12} (two last rows show the RGB and depth key-frames).

left hand. The z-axis is latent with body. The origin coordinate is the starting point when subject rises hand in front of their body and pointing forward to the Kinect sensor K_3.

We follow the evaluation protocol presented in our previous research [2]. The protocol is leave-one-subject-out cross-validation. That means samples from each subject are used once for testing while samples from the remaining subjects form the training set. We then compute the average accuracy of every holdout.

4.2   Experimental Results

In this section we will report the recognition results obtained by our proposed framework. The experiments will be conducted on five viewpoints. We compare the use of single modality with combination of multiple modalities (RGB, Optical flow and Depth). The accuracies of different fusion methods (early fusion, late fusion and multiple kernel learning) will be reported.

Table 1 shows accuracies obtained by using sole RGB, Optical flow, Depth or combination of RGB, Optical flow and Depth with different fusion techniques. In this table, we also show the kernel functions that we used for each MKL and corresponding SVM. Table 2 extracts the average accuracy obtained from five Kinect viewpoints. We have some observations as follows:

Comparison Between Modalities. As rbf kernel gives the best average accuracy on five Kinects views, we will analyze results using this kernel function.

- RGB gives better performance than depth. Using rbf kernel function, accuracy obtained by RGB stream is higher than the one obtained by Depth stream. The reason is that Depth is quite noisy. Especially when hand is of low resolution, the quality of Depth on hand region is very poor. Some parts of hand

Table 1. Accuracy (%) comparison of methods using single or multiple modalities with different fusion techniques

View	Kernel	RGB	OF	Depth	RGB-OF			RGB-Depth			OF-Depth			RGB-OF-Depth		
					Early	Late	MKL	Early	Late	MKL	Early	Late	MKL	Early	Late	MKL
1	Laplacian	67.0	74.3	66.9	**77.8**	73.8	75.6	70.6	67.7	69.1	74.5	72.5	73.1	75.9	71.6	74.4
	Linear	67.6	73.8	64.4	**77.5**	73.0	74.3	69.4	66.6	67.9	73.9	70.5	72.6	75.9	71.1	74.0
	Rbf	68.8	73.4	65.4	**77.7**	73.0	74.5	69.9	66.3	68.6	74.4	71.1	71.6	75.4	72.2	73.9
2	Laplacian	78.2	76.3	71.4	79.4	77.5	78.8	78.7	74.3	**79.7**	79.5	76.1	78.2	79.5	78.8	79.2
	Linear	78.1	76.0	73.4	79.4	77.4	80.2	79.7	77.3	**81.1**	79.1	77.1	77.9	80.4	79.1	80.3
	Rbf	78.1	76.5	71.9	79.4	77.8	77.8	79.1	76.4	80.8	79.0	77.1	76.7	80.1	78.2	**81.0**
3	Laplacian	86.3	82.3	71.3	**88.6**	88.3	87.9	86.7	85.3	86.9	84.5	82.7	83.3	87.7	86.3	88.0
	Linear	85.9	82.3	68.1	**88.2**	88.2	87.8	86.9	83.5	87.2	83.3	80.2	83.3	87.9	85.4	88.0
	Rbf	85.4	82.7	69.8	**88.2**	87.9	87.5	86.9	83.9	86.9	83.7	81.3	82.1	88.0	85.4	87.5
4	Laplacian	80.7	79.7	58.9	82.1	**83.0**	81.6	80.0	75.9	80.0	80.3	74.6	78.8	83.0	80.8	81.6
	Linear	80.5	80.5	54.4	81.1	82.6	81.1	80.0	73.1	80.1	80.4	74.3	79.3	**83.4**	81.3	81.3
	Rbf	80.0	80.0	56.9	81.4	82.4	82.2	79.8	73.4	78.3	80.4	75.7	80.0	**83.4**	81.5	82.7
5	Laplacian	67.1	75.4	61.7	72.6	71.7	73.8	69.4	67.5	68.5	76.0	74.8	75.7	**75.7**	73.3	73.0
	Linear	65.9	74.2	60.9	72.1	72.3	72.6	69.1	67.3	65.1	75.1	74.0	**75.2**	75.1	72.3	72.5
	Rbf	67.2	74.8	62.6	72.1	72.6	72.1	68.6	67.4	70.2	75.2	74.8	**75.8**	75.5	73.1	74.4

Table 2. Average accuracy (%) on five viewpoints

Kernel	RGB	OF	Depth	RGB-OF			RGB-Depth			OF-Depth			RGB-OF-Depth		
				Late	Early	MKL	Late	Early	MKL	Late	Early	MKL	Eary	Late	MKL
Laplacian	75.9	77.6	66.0	80.1	78.9	79.5	77.1	74.1	76.9	79.0	76.2	77.8	**80.4**	78.2	79.3
Linear	75.6	77.4	64.2	79.7	78.7	79.2	77.0	73.6	76.3	78.4	75.2	77.7	**80.5**	77.8	79.2
Rbf	75.9	77.5	65.3	79.8	78.7	78.8	76.9	73.5	77.0	78.5	76.0	77.2	**80.5**	78.1	79.9

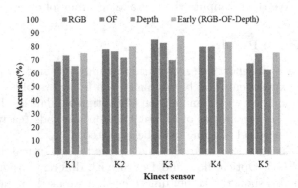

Fig. 6. Comparison of the best accuracy obtained by single or multiple modalities using early fusion. The horizontal axis represents Kinect views. The vertical axis represents accuracy (%) for those views.

or fingers are missing on Depth data. At difficult views (K_1, K_5), Depth accuracy is lower than RGB accuracy by 3.34% on K_1, 4.26% on K_5. However, at frontal or almost frontal views, Depth is significantly worse than RGB on three Kinect sensors K_2, K_3, K_4 with 6.24%, 15.6% and 23.04% respectively.

This observation is different w.r.t. [8] which confirmed that Depth is much better than RGB thanks to the high resolution of the hand on the frame.
- Optical flow is not better than RGB at frontal views (K_2, K_3, K_4). However, Optical flow outperforms RGB at the hardest views (exceeding by 4.6% on K_1 and 7.6% on K_5 if using rbf kernel function). It means that at difficult views, the hand posture could cause confusion to the recognizer while hand loco-motion helps at clearer discriminating. Optical flow is always more authentic than Depth.

Comparison of Single Modality with Multiple Modalities

- Combination of two modalities or three modalities always helps increasing the performance of recognition. The highest accuracy was obtained with the combination of all three modalities (RGB, Depth, Optical flow). In average of 5 Kinect views, accuracies are 80.5% by early fusion, 78.1% by late fusion and 80.1% by MKL. Early fusion gives the best results and slightly higher than MKL. Therefore, we could choose early fusion for computation simplification.
- Combination of three modalities is only slightly better (80.5%) than combi-nation of RGB with Optical flow (79.8%) using early fusion. Therefore, one recommendation that we can just combine RGB with Optical flow to reduce computational time.
- Among three fusion techniques, late fusion is the least efficient in almost every case. The reason could be when Depth is not good, decision based on scores can not improve. We have tried max scores fusion and average score fusion and found that average score fusion is generally better than max score fusion. The result reported in this paper is average score fusion.
- Viewpoint has strong impact on recognition accuracy (Fig. 6). For the frontal view (K_3), the highest accuracy is obtained with MKL and early fusion (88.0%). We can observe fully and easily hand gestures from this viewpoint. The accuracy at two neighbouring views of K_2, K_4 reduces to 80.4% and 83.4%. The lowest accuracy (75.9% and 75.7%) obtained at K_1 and K_5 because hands are largely occluded and it is difficult to observe the whole gesture from these perspectives.

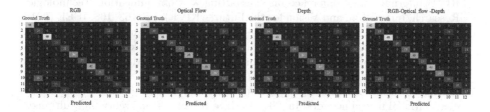

Fig. 7. Confusion matrices obtained by using single RGB, Depth, Optical Flow or their combination with Early Fusion.

Figure 7 shows the confusion matrices obtained by using single RGB, Depth or Optical Flow or their combination with early fusion at the most difficult view K_5. We observe that in term of hand kinetic of gesture G_5 and G_{12} are very similar, now better recognized by RGB-OF-Depth than using uni-modality.

5   Conclusions

In this paper, we have presented a new framework for human hand gesture recognition using multi-modal data streams. The framework utilized 3D convolutional neural networks for extracting spatial-temporal features from RGB, optical flow and depth sequence. These features are subsequently inputted into a fusion module for classification. The highest accuracy obtained with frontal view (88.0%) while the accuracy gradually reduced to 80% or 75% on the side views. In the near future, we will explore other modalities and try other feature extractors for better capture discriminant features for recognition. We also evaluate the proposed framework on other benchmark datasets and deploy the proposed method in a human machine interaction application.

References

1. Aiolli, F., Da San Martino, G., Sperduti, A.: A kernel method for the optimization of the margin distribution. In: Kůrková, V., Neruda, R., Koutník, J. (eds.) ICANN 2008. LNCS, vol. 5163, pp. 305–314. Springer, Heidelberg (2008). https://doi.org/10.1007/978-3-540-87536-9_32
2. Dang, M.T., Doan, H.G., Tran, T.H., Le, T.L., Vu, H.: Robustness analysis of 3D convolutional neural network for human hand gesture recognition. Int. J. Mach. Learn. Comput. **9**(2), 135–142 (2019)
3. Devineau, G., Moutarde, F., Xi, W., Yang, J.: Deep learning for hand gesture recognition on skeletal data. In: 2018 13th IEEE International Conference on Automatic Face Gesture Recognition (FG 2018), pp. 106–113 (2018)
4. Doan, H.G., Nguyen, V.T., Vu, H., Tran, T.H.: A combination of user-guide scheme and kernel descriptor on RGB-D data for robust and realtime hand posture recognition. Eng. Appl. Artif. Intell. **49**, 103–113 (2016)
5. Doan, H.G., Vu, H., Tran, T.H.: Phase synchronization in a manifold space for recognizing dynamic hand gestures from periodic image sequence. In: 2016 IEEE RIVF International Conference on Computing & Communication Technologies, Research, Innovation, and Vision for the Future (RIVF), pp. 163–168. IEEE (2016)
6. Hoang, N.N., Lee, G.S., Kim, S.H., Yang, H.J.: A real-time multimodal hand gesture recognition via 3D convolutional neural network and key frame extraction. In: Proceedings of the 2018 International Conference on Machine Learning and Machine Intelligence, pp. 32–37. ACM (2018)
7. Khong, V.M., Tran, T.H.: Improving human action recognition with two-stream 3D convolutional neural network. In: 2018 1st International Conference on Multimedia Analysis and Pattern Recognition (MAPR), pp. 1–6. IEEE (2018)
8. Köpüklü, O., Gunduz, A., Kose, N., Rigoll, G.: Real-time hand gesture detection and classification using convolutional neural networks. CoRR abs/1901.10323 (2019)

9. Kurakin, A., Zhang, Z., Liu, Z.: A real time system for dynamic hand gesture recognition with a depth sensor. In: 2012 Proceedings of the 20th European signal processing conference (EUSIPCO), pp. 1975–1979. IEEE (2012)
10. Liu, K., Zhou, F., Wang, H., Fei, M., Du, D.: Dynamic hand gesture recognition based on the three-dimensional spatial trajectory feature and hidden Markov model. In: Li, K., Fei, M., Du, D., Yang, Z., Yang, D. (eds.) ICSEE/IMIOT -2018. CCIS, vol. 924, pp. 555–564. Springer, Singapore (2018). https://doi.org/10.1007/978-981-13-2384-3_52
11. Narayana, P., Beveridge, J.R., Draper, B.A.: Gesture recognition: focus on the hands. In: Proceedings of the 2018 International Conference on Pattern Recognition, pp. 5235–5234 (2018)
12. Ofli, F., Chaudhry, R., Kurillo, G., Vidal, R., Bajcsy, R.: Berkeley MHAD: a comprehensive multimodal human action database. In: 2013 IEEE Workshop on Applications of Computer Vision (WACV), pp. 53–60. IEEE (2013)
13. Ohn-Bar, E., Trivedi, M.M.: Hand gesture recognition in real time for automotive interfaces: a multimodal vision-based approach and evaluations. IEEE Trans. Intell. Transp. Syst. **15**(6), 2368–2377 (2014)
14. Rautaray, S.S., Agrawal, A.: Vision based hand gesture recognition for human computer interaction: a survey. Artif. Intell. Rev. **43**(1), 1–54 (2015)
15. Ren, Z., Meng, J., Yuan, J.: Depth camera based hand gesture recognition and its applications in human-computer-interaction. In: 2011 8th International Conference on Information, Communications & Signal Processing, pp. 1–5. IEEE (2011)
16. Ren, Z., Yuan, J., Zhang, Z.: Robust hand gesture recognition based on finger-earth mover's distance with a commodity depth camera. In: Proceedings of the 19th ACM International Conference on Multimedia, pp. 1093–1096. ACM (2011)
17. Simonyan, K., Zisserman, A.: Two-stream convolutional networks for action recognition in videos. In: Advances in Neural Information Processing Systems, pp. 568–576 (2014)
18. Tran, D., Bourdev, L., Fergus, R., Torresani, L., Paluri, M.: Learning spatiotemporal features with 3D convolutional networks. In: Proceedings of the IEEE International Conference on Computer Vision, pp. 4489–4497 (2015)
19. Wu, Q., Wang, Z., Deng, F., Chi, Z., Feng, D.D.: Realistic human action recognition with multimodal feature selection and fusion. IEEE Trans. Syst. Man Cybern. Syst. **43**(4), 875–885 (2013)
20. Zhang, C., Tian, Y.: Edge enhanced depth motion map for dynamic hand gesture recognition. In: 2013 IEEE Conference on Computer Vision and Pattern Recognition Workshops, pp. 500–505, June 2013
21. Zhu, G., Zhang, L., Shen, P., Song, J.: Multimodal gesture recognition using 3-D convolution and convolutional LSTM. IEEE Access **5**, 4517–4524 (2017)

Cross-Domain Face Recognition
Using Dictionary Learning

Yaswanth Gavini[1(✉)], Arun Agarwal[1], and B. M. Mehtre[2]

[1] School of Computer and Information Sciences, University of Hyderabad,
Hyderabad, India
`yaswanth.gavini@uohyd.ac.in`, `aruncs.2011@gmail.com`
[2] Centre of Excellence in Cyber Security, Institute for Development and Research
in Banking Technology, Hyderabad, India
`bmmehtre@idrbt.ac.in`

Abstract. Cross-domain face recognition refers to the matching of face images between different domains. It has many applications in night time surveillance, border security surveillance and law-enforcement. However, this is a difficult task because of the non-linear intensity pixel values between the images which occur due to the domain gap. Recently, dictionary learning methods such as coupled dictionary learning and domain adaptive dictionary learning methods are used to solve this problem. In this paper, we propose a dictionary learning based method to learn the common subspace in order to reduce the gap between domains. Initially, we separate the domain specific representation and identity related representation by using commonality and particularity dictionary learning. In the next step, we remove the domain specific representation and get the common subspace. Thereafter, in order to get the more discriminate representation, we use metric learning. The proposed method is tested on RGB-D-T data set and the experimental results show that the proposed method is performing better even when there is no person common between training and testing sets.

Keywords: Cross-domain · Face recognition · Common subspace

1 Introduction

In recent years, there is a significant advancement in the surveillance system. One of the key capabilities of the surveillance system is low light or night time surveillance. For this capability, cameras use spectral bands other than visual spectrum. Using the visual spectrum in night time surveillance is impractical as visual spectrum always need an external illumination source. Hence, NIR (near-infrared) and thermal bands are utilized for the night time surveillance. In automatic surveillance systems, it is essential to match the night time captured data with day time captured data, which enables a new challenge of cross-domain face recognition. Cross-domain face recognition is a challenging task as the spectral characteristics of face are different in visual spectrum and other (NIR and

R. Chamchong and K. W. Wong (Eds.): MIWAI 2019, LNAI 11909, pp. 168–180, 2019.
https://doi.org/10.1007/978-3-030-33709-4_15

thermal). These characteristics result in non-linear pixel intensity values between the domains.

Cross-domain face recognition is not only used in automatic surveillance but also utilized in applications of law enforcement. The task of cross-domain face recognition is to actively match between sketch to visual, NIR to visual, and thermal to visual face images. At times, cross-domain face recognition is also referred to as heterogeneous face recognition (HFR). There are three categories [13] in solving the cross-domain face recognition, namely feature based, synthesis based, and projection based approaches. Feature based approaches [3] try to select and extract features which are domain invariant. In synthesis based methods [27], cross-domain face recognition is done in two stages. In the first stage, it tries to project one domain images to another domain (usually target domain is visual, source domain is any one of the sketch or NIR or thermal). It takes any of the source domain face image as input and tries to predict the corresponding image in visual domain (target). In the second stage it uses the single domain face recognition as all are projected into a single domain. In projection based approaches [17], both domain images get projected into a common subspace where they are more comparable than original space.

Dictionary learning is a widely used approach in signal processing applications like compressive sensing, signal denoising, image super resolution and signal classification. Any signal can be expressed by the linear combination of basis vectors and this set of basis vectors/atoms is called a dictionary. Dictionary learning is termed as sparse coding or sparse representation as usually any signal is expressed by a linear combination of few basis vectors. Number of atoms and sparsity level plays a key role in dictionary learning [8]. Dictionary learning is used for face recognition, and the dictionary learning for face recognition is classified into five categories [24] namely - shared dictionary learning, class specific dictionary learning, auxiliary dictionary learning, commonality and particularity dictionary learning, and domain adaptive dictionary learning. In shared dictionary learning [4,7], there exists one shared dictionary for all the classes, whereas in class specific dictionary learning [1,26] each class has its own dictionary. Auxiliary dictionary learning [23] is used when the availability of training data is less. In commonality and particularity dictionary learning [22,25] there are two dictionaries - particularity dictionary and commonality dictionary. Class specific features are present in particularity dictionary, whereas common features which exist in all the classes are present in commonality dictionary. Domain adaptive dictionary learning [14,28] is used when the training data and testing data are in different distributions (e.g., cross-domain face recognition).

Metric learning is a branch of machine learning which learns the function to estimate the similarity/distance between the two objects. The function will be learned from the input data. When compared to the conventional distance metrics, metric learning methods [9] are more robust as the similarity gets learned from the data.

In this paper, we have proposed a dictionary learning based cross-domain (thermal to visual) face recognition which has two stages. In the first stage, we project both domain images into a common subspace in which the face images are represented with sparse code. In the second stage, metric learning is done for measuring the similarity between corresponding sparse codes. In order to find the common subspace, we use commonality and particularity dictionary learning. Here, we have considered all visual face images as one class and all thermal face images as another class. By using commonality and particularity dictionary learning, we separate commonality features between the classes (visual and thermal), and particularity between classes. Here the same person face image is present in both the classes (domains) and the common features between the two classes are person identity related features. These identity related features are present in commonality dictionary, whereas domain specific features are present in particularity dictionary. On the other hand, in conventional single domain commonality and particularity dictionary learning, identity related features are present in particularity dictionary. By removing domain specific features from the input data, we get common subspace in which the person identity is preserved. After projecting into common subspace, the common subspace representation is learned with commonality dictionary. This learned representation is used to find the similarity between two images, where each image is represented with one sparse code. In order to estimate the similarity between sparse codes, we use large scale metric learning (LSML) [9].

The rest of the paper is organized as follows. The related work is given in Sect. 2. Section 3 discusses the proposed method. Experimental results and analysis are discussed in Sect. 4, and the conclusion and future work are in Sect. 5.

2   Related Work

Most of the early research methods of cross-domain face recognition are related to handcrafted features which is one of the feature based methods. Hu *et al.* [5] proposed a method which extracts the handcrafted features like SIFT, HOG, and LBP for both visual and thermal domain images and matches the extracted features using partial least square discriminate analysis (PLS-DA). Sarfraz *et al.* [19] proposed a method in which the SIFT features of both thermal and visual domain images get extracted and the obtained features get matched using deep neural networks. The other category of feature based methods are automatic feature learning methods. There are mainly two approaches in automatic feature learning methods namely deep learning based methods and dictionary learning based methods. Riggan *et al.* [18] proposed neural network based coupled architecture network which forces the hidden layers of two neural networks to be as similar as possible. Gavini *et al.* [2] proposed a convolution neural network (CNN) based transfer learning method in which the visual features are learned via CNN, and the learnt features get transferred to the thermal domain. Reale *et al.* [15] proposed a variant of contrastive loss to train the convolution neural network based coupled network. Coupled dictionary learning algorithms are proposed in [6,16], which learns the shared feature space for cross-domain image data.

In our work, we consider the commonality and particularity dictionary learning [20,21], which is used for single domain face recognition algorithms. In these algorithms, for each class a separate dictionary is learned which is similar to class specific dictionary learning. In addition to the class specific dictionary, it also learns the shared dictionary. Shared dictionary learns the common features which are present in all the classes. These common features are used only for the representation. By separating common features from class specific features, class specific features discrimination capability gets improved.

Mudunuri et al. [11] proposed a dictionary aligned low resolution and heterogeneous face recognition method. This method first learns the orthogonal dictionary for two domains and align the dictionary atoms based on atom correlation. With these aligned dictionaries the method computes the aligned sparse representation. To find the final similarity, the authors have used metric learning on the sparse representation. In our method, by projecting the data into the common subspace we get the aligned sparse representation.

3   Proposed Method

This section describes the proposed method of thermal to visual face recognition, which is a two stage approach. In the first stage, we find the common subspace, and we also get the sparse representation for reduced space. This reduced sparse representation may not be discriminant. To get the discriminant sparse representation, we use metric learning. Let the set of visual domain images be $X_v \in \Re^{d \times n_v}$, where n_v is the number of visual domain images, d is the dimensionality of image and $X_t \in \Re^{d \times n_t}$ be the set of thermal domain images, where n_t is the number of thermal domain images. Let $X \in \Re^{d \times N}$ represent both visual and thermal image data, $X = [X_v \ X_t]$. Here, N represents the total number of visual and thermal images i.e., $N = n_v + n_t$. On the other hand, let $\hat{D} \in \Re^{d \times \hat{k}}$ be the overall dictionary having two parts D and D_0. Column wise concatenation of D and D_0 is $\hat{D}$ i.e., $\hat{D} = [D \ D_0]$. Here, $\hat{k} = k_v + k_t + k_0$ is the dimensionality of latent space for whole data, k_0 is the dimensionality of latent space for common data, k_t is the dimensionality of latent space of thermal data and k_v is the dimensionality of latent space of visual data. D_0 is the shared dictionary between two domains and the atoms of D_0 represent the commonality between the two domains. Here, same person face is represented in two domains, so the person identity related features are common between the two domains and so $D_0 \in \Re^{d \times k_0}$ represents person identity features. Let $D_t \in \Re^{d \times k_t}$ be the thermal domain specific dictionary and $D_v \in \Re^{d \times k_v}$ be the visual domain specific dictionary. Let $D \in \Re^{d \times k}$ is the domain specific combined dictionary and the column concatenation of D_v and D_t is D i.e., $D = [D_v \ D_t]$. Here $k = k_v + k_t$ is the dimensionality of latent space for class specific data.

Let $\alpha_t \in \Re^{k_t \times N}$ represents the thermal domain sparse representation and is learned with the thermal class specific dictionary D_t, and $\alpha_v \in \Re^{k_v \times N}$ represent visual domain sparse representation, which will be learned with visual domain dictionary D_v. The domain specific sparse representation is $\alpha \in \Re^{k \times N}$ and

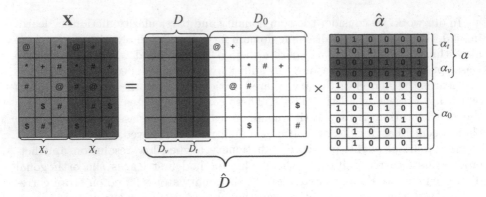

Fig. 1. Illustration of input data (X) in terms of dictionary $(\hat{D})$ and sparse code $(\hat{\alpha})$. Let the colour represents domain specific features and $\{@, *, \#, \$, +\}$ represent identity related features. (Color figure online)

$\alpha = [\alpha_v \ \alpha_t]^T$. Let $\alpha_0 \in \Re^{k_0 \times N}$ represent commonality sparse representation, which will be learned with shared dictionary D_0. Combined sparse representation is $\hat{\alpha} = [\alpha \ \alpha_0]^T$. Let m, m_0, m_v, m_t be the mean vectors of $\alpha, \alpha_0, \alpha_v, \alpha_t$ respectively.

Figure 1 illustrates the combined dictionary $\hat{D}$ and sparse representation $\hat{\alpha}$. X has two parts - X_v and X_t. Each column of X_v represents the visual face image and each column of X_t represents thermal face image. Let us assume that the colour represents the domain specific features and the set of symbols $\{@, *, \#, \$, +\}$ represent the identity related features. Our goal is to separate the domain specific features and identity related features. The domain specific features are in dictionary D and the identity related features are in dictionary D_0.

3.1   Common Subspace Learning

Common subspace learning (CSL) is very important, and the first stage in the proposed method. In this stage, we separate the domain specific representation features and identity related features. We have proposed two common subspace learning methods namely CSL1 and CSL2. Both of the methods try to separate the domain specific and identity related atoms and learn the corresponding sparse representation. Let $x^i \in X$, and for each x^i the corresponding sparse representation is $\hat{\alpha}^i$ as $x^i = \hat{D}\hat{\alpha}^i$, where $\hat{D} = [D \ D_0], \hat{\alpha} = [\alpha \ \alpha_0]$. It can be rewritten as $x^i = D\alpha^i + D_0\alpha_0^i$. $D\alpha^i$ is the domain specific representation and $D_0\alpha_0^i$ is the common subspace representation for x^i. For the cross domain matching, domain specific representation $(D\alpha^i)$ is not needed, and so we discard the domain specific part from x^i and consider the remaining part as the new common subspace for x^i, i.e., $D_0\alpha_0^i$. Hence the corresponding common subspace sparse code for x^i is α_0^i. The output of both the methods of common subspace learning is α_0.

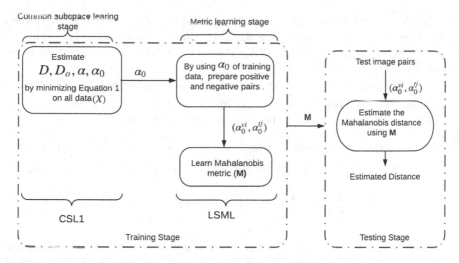

Fig. 2. Illustration of proposed method-1 (CSL1+LSML). CSL1 is used for common subspace learning and for metric learning, LSML [9] is used.

Common Subspace Learning Method1 (CSL1): The input to this stage is X and the domain labels. CSL1 uses domain labels only. At this stage, we learn D, α, D_0 and α_0 by minimizing the following objective function (Eq. 1).

$$J(D, \alpha, D_0, \alpha_0) = ||X - D_0\alpha_0 - D\alpha||_F^2 + ||X_t - D_0\alpha_0^t - D_t\alpha_t^t||_F^2 + ||D_v\alpha_v^t||_F^2$$
$$+ ||X_v - D_0\alpha_0^v - D_v\alpha_v^v||_F^2 + ||D_t\alpha_t^v||_F^2 + \lambda_1||\alpha||_1 + \lambda_1||\alpha_0||_1 + \lambda_2 f(\hat{\alpha}) \tag{1}$$

where λ_1, λ_2 are regularization parameters and,

$$f(\hat{\alpha}) = ||\alpha_v - M_v||_F^2 + ||\alpha_t - M_t||_F^2 - ||M_v - M||_F^2 - ||M_t - M||_F^2$$
$$+ ||\alpha_0 - M_0||_F^2 + ||\alpha||_F^2 \tag{2}$$

The objective function (Eq. 1) is not jointly convex in D, α, D_0, α_0. But it is convex separably w.r.t each of the D, α, D_0, α_0. So an algorithm that alternatively updates each variable by fixing the others can be used. The updation of α, α_0 and D can be done in a similar way as that of LRSDL [21]. For D_0, in LRSDL it tries to put in low rank, whereas in our problem D_0 is in higher rank as the identity related atoms are in D_0. So the updation of D_0 is done using online dictionary learning [10]. Illustration of the proposed method shown in Fig. 2.

Common Subspace Learning Method2 (CSL2): The main difference between CSL1 and CSL2 is that the CSL1 doesn't use any identity specific labels but uses only domain related labels. CSL2 uses identity related features and domain related labels too. CSL2 has three stages as shown in Fig. 3. The first and third stages are similar to that of CSL1, whereas the Stage-2 of CSL2 uses only the identity related labels. The main idea of CSL2 is to learn more refined domain specific features which are learned by using identity related labels. The purpose of Stage-1 and Stage-3 in CSL2 is to get the dictionary aligned common subspace sparse code for all images in X.

Let X^l, X_t^l and X_v^l be the identity labeled training data, and $\alpha_0^{v/i}$ be the commonality sparse code for the visual image of class i, and similarly $\alpha_0^{t/i}$ be the commonality sparse code for the thermal image of class i.

$$
\begin{aligned}
J(D, \alpha, D_0, \alpha_0) =& ||X^l - D_0\alpha_0 - D\alpha||_F^2 + ||X_t^l - D_0\alpha_0^t - D_t\alpha_t^t||_F^2 + ||D_v\alpha_v^t||_F^2 \\
&+ \lambda_1||\alpha||_1 + \lambda_1||\alpha_0||_1 + ||X_v^l - D_0\alpha_0^v - D_v\alpha_v^v||_F^2 + ||D_t\alpha_t^v||_F^2 \\
&+ \sum_{i=1}^{C} ||\alpha_0^{v/i} - \alpha_0^{t/i}||_F^2 + \lambda_2 f(\hat{\alpha})
\end{aligned}
$$

$$(3)$$

In Stage-1 of CSL2, we estimate D, α, D_0 and α_0 by minimizing the objective function (Eq. 1). Thereafter the D and D_0 of Stage-2 are initialised with learned D and D_0 of Stage-1 and then update D, α, D_0 and α_0 by minimizing the objective function (Eq. 3). The learned D and D_0 of Stage-2 get transferred to Stage-3 of CSL2 i.e., D and D_0 of Stage-3 are initialised with learned D and D_0 of Stage-2. In Stage-3, we fix D and update α, D_0 and α_0 by minimizing the objective function (Eq. 3). The updation of α and D can be done in a similar way as that of LRSDL [21]. Updation of α_0 varies slightly because of the additional term $||\alpha_0^{v/i} - \alpha_0^{t/i}||_F^2$. Updation of D_0 is done using online dictionary learning [10].

3.2   Metric Learning

After completion of common subspace learning, we get common subspace sparse code α_0 for all data X. This sparse code α_0 is not inherently discriminative, and to make it discriminative, we use large scale metric learning (LSML) [9]. In this, we estimate the log-likelihood between learned sparse codes and for that we need to get the distance between sparse codes as given below,

$$
Dist(\alpha_0^{vi}, \alpha_0^{tj}) = (\alpha_0^{vi} - \alpha_0^{tj})M(\alpha_0^{vi} - \alpha_0^{tj})^T
$$

Here M is the Mahalanobis metric and the vector difference between the sparse code is formulated as given below,

$$
Diff(\alpha_0^{vi}, \alpha_0^{tj}) = log\left(\frac{\frac{1}{\sqrt{2\pi|\Sigma_{z_{ij}=0|}}} exp^{(-1/2(\alpha_0^{vi}-\alpha_0^{tj})^T \Sigma_{z_{ij}=0}^{-1} (\alpha_0^{vi}-\alpha_0^{tj}))}}{\frac{1}{\sqrt{2\pi|\Sigma_{z_{ij}=1|}}} exp^{(-1/2(\alpha_0^{vi}-\alpha_0^{tj})^T \Sigma_{z_{ij}=1}^{-1} (\alpha_0^{vi}-\alpha_0^{tj}))}} \right)
$$

where,

$$
\Sigma_{z_{ij}=0} = \sum_{z_{ij}=0} (\alpha_0^{vi} - \alpha_0^{tj})(\alpha_0^{vi} - \alpha_0^{tj})^T
$$

$$
\Sigma_{z_{ij}=1} = \sum_{z_{ij}=1} (\alpha_0^{vi} - \alpha_0^{tj})(\alpha_0^{vi} - \alpha_0^{tj})^T
$$

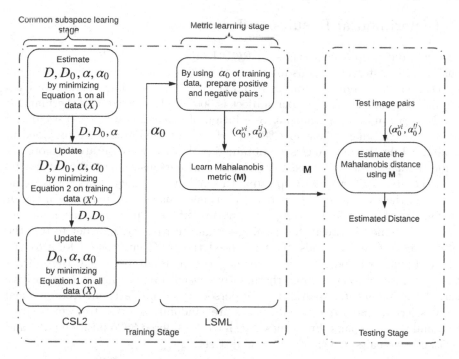

Fig. 3. Illustration of proposed method-2 (CSL2+LSML). For common subspace learning, CSL2 is used, and for metric learning, LSML [9] is used.

Here, z is an indicator matrix of size $n_v \times n_t$. The term $z_{ij} = 0$ indicates that α_0^{vi} and α_0^{tj} are of same person's sparse codes (positive pair) and, $z_{ij} = 1$ indicates that α_0^{vi} and α_0^{tj} are of different persons sparse codes (negative pair). By solving the above equations we obtain Mahalanobis distance metric M and the detailed algorithm is in [9].

In this method, testing is done in the following way. Let x_v^i, x_t^j be the test images of visual and thermal face images. By utilising the learned dictionaries of common subspace learning D and D_0, we find the corresponding sparse representation α_0^{vi}, α_0^{tj}. By using the learned distance metric M, we estimate the similarity between both the images, and based on the similarity we decide whether both the images are similar are not.

CSL1 + LSML, CSL2 + LSML are our proposed methods. For simplicity of notation, we use Method-1 for CSL1 + LSML, and Method-2 for CSL2 + LSML. In Method-1, first it learns the common subspace using CSL1 method and then the large scale metric learning (LSML) [9] is used. Similarly, in Method-2 common subspace learning is done by CSL2 method, and then large scale metric learning is used.

4   Experimental Results and Analysis

We tested our proposed methods on RGB-D-T dataset [12]. RGB-D-T dataset consists of 51 different persons images in three different domains - visual, depth and thermal. Each domain contains 15,300 (51 × 300) face images, and for each domain the number of images per subject is 300. We have used only thermal and visual images in our experiments. Resolution of visual images is 640 × 480 and that of thermal images is 384 × 288. The details of the dataset are provided in Table 1. All the thermal images are obtained in the long wave infrared region. We divide the dataset which consists of 51 different persons images into two parts. From the first part, we use 26 persons thermal-visual image pairs for training and, from the second part 25 persons thermal-visual image pairs are used for testing. Training set and testing set contains different persons images and no person is common between the training set and testing set. In order to test the performance of the methods, we have used the ROC curve, and for this the balanced number of positive pairs and negative pairs are preferred. Here, if the same person is present in both thermal and visual images then it is a positive pair, else it is a negative pair. For each person in second part, one frontal visual face is selected, and 153 thermal images are randomly selected to form positive pairs and negative pairs. From these, we have considered 3825 positive pairs and 3825 negative pairs, which are used as test set.

Table 1. Details of RGB-D-T dataset [12]

Number of subjects	Number of visual images	Number of thermal images	Resolution of visual images	Resolution of thermal images
51	15,300	15,300	640 × 480	384 × 288

Table 2. Experimental results on RGB-D-T data

Method	EER	AUC
PLS-DA [5]	0.334	0.738
CSL1+LSML (Method-1)	0.335	0.716
CSL2+LSML (Method-2)	**0.169**	**0.907**

The experimental results on RGB-D-T dataset are shown in Table 2. Column-1 shows the methods which we have experimented, whereas second and third columns give the EER (equal error rate) and AUC (area under receiver operating characteristics curve) metric values. Lower the value of EER, higher the value of AUC, better the performance of the method. Our proposed methods are compared with the baseline algorithm - PLS-DA [5]. By looking at the

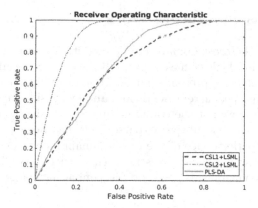

Fig. 4. ROC curve

results, we can say that Method-2 (our method) is performing better, and the performance of PLS-DA and Method-1 are very close.

Figure 4 shows the ROC curve of all methods. By observing the figure, we can say that Method-2 is performing better. Also, the closeness of PLS-DA and Method-1 can be observed from the figure.

For training common subspace learning, dictionary size plays an important role. We need to select the size in such a way that the size of the common dictionary (k_0) is greater than that of the size of domain specific dictionary (k). Reason for this is, domain specific dictionary atoms contribute more for representation not for discrimination. Redundancy between atoms is high, so the rank of the domain specific dictionary is low. We have experimented our method with different dictionary sizes and different learning rates. Table 3 shows the parameter values for which we got the best results. For CSL1, dictionary size ($\hat{k}$) is 340 ($k_t + k_v + k_0$) and for CSL2, the dictionary size ($\hat{k}$) is 460 ($k_t + k_v + k_0$). The learning rate in metric learning after parameter tuning is set to 0.1 for both the methods. To implement CSL1 and CSL2, we have used discriminating dictionary learning library toolbox [20, 21].

Table 3. Parameter values (through which the best results are achieved) used in the proposed method

Method	Dictionary sizes				Regularization parameters		
	k_t	k_v	k_0	$\hat{k}$	λ_1	λ_2	λ_3
CSL1	20	20	300	340	0.001	0.001	NA
CSL2	30	30	400	460	0.001	0.001	0.002

5   Conclusions and Future Work

We have proposed the cross-domain face recognition between thermal domain and visual domain, in which we have separated the domain specific representation from the identity related representation. We tried to reduce the domain gap by removing the domain specific representation, and by considering only the identity related representation we got the common subspace. Using metric learning, the discrimination in common subspace sparse code has been improved which in turn improved the performance of the cross-domain face recognition.

In the future, we would like to investigate the kernel based dictionary learning in order to get the common subspace and discriminating sparse code at the same time.

Acknowledgements. This work has been supported by Ministry of Electronics & Information Technology (MeitY), Government of India under the project Visvesvaraya PhD Scheme, which is implemented by Digital India Corporation (formerly Media Lab Asia).

The authors would also like to acknowledge the funding support from DST PURSE-II, Government of India for the high performance computing facility.

References

1. Cai, S., Zuo, W., Zhang, L., Feng, X., Wang, P.: Support vector guided dictionary learning. In: Fleet, D., Pajdla, T., Schiele, B., Tuytelaars, T. (eds.) ECCV 2014. LNCS, vol. 8692, pp. 624–639. Springer, Cham (2014). https://doi.org/10.1007/978-3-319-10593-2_41
2. Gavini, Y., Mehtre, B.M., Agarwal, A.: Thermal to visual face recognition using transfer learning. In: 2019 IEEE 5th International Conference on Identity, Security, and Behavior Analysis (ISBA), pp. 1–8, January 2019. https://doi.org/10.1109/ISBA.2019.8778474
3. Gong, D., Li, Z., Huang, W., Li, X., Tao, D.: Heterogeneous face recognition: a common encoding feature discriminant approach. Trans. Image Process. **26**(5), 2079–2089 (2017). https://doi.org/10.1109/TIP.2017.2651380
4. Guo, H., Jiang, Z., Davis, L.S.: Discriminative dictionary learning with pairwise constraints. In: Lee, K.M., Matsushita, Y., Rehg, J.M., Hu, Z. (eds.) ACCV 2012. LNCS, vol. 7724, pp. 328–342. Springer, Heidelberg (2013). https://doi.org/10.1007/978-3-642-37331-2_25
5. Hu, S., Choi, J., Chan, A.L., Schwartz, W.R.: Thermal-to-visible face recognition using partial least squares. J. Opt. Soc. Am. A **32**(3), 431–442 (2015). https://doi.org/10.1364/JOSAA.32.000431
6. Huang, D., Wang, Y.F.: Coupled dictionary and feature space learning with applications to cross-domain image synthesis and recognition. In: 2013 IEEE International Conference on Computer Vision, pp. 2496–2503, December 2013. https://doi.org/10.1109/ICCV.2013.310
7. Jiang, Z., Lin, Z., Davis, L.S.: Learning a discriminative dictionary for sparse coding via label consistent k-svd. In: CVPR 2011, pp. 1697–1704 (June 2011). https://doi.org/10.1109/CVPR.2011.5995354

8. Rajesh, K., Negi, A.: Heuristic based learning of parameters for dictionaries in sparse representations. In: 2018 IEEE Symposium Series on Computational Intelligence (SSCI), pp. 1013–1019, November 2018. https://doi.org/10.1109/SSCI.2018.8628661

9. Köstinger, M., Hirzer, M., Wohlhart, P., Roth, P.M., Bischof, H.: Large scale metric learning from equivalence constraints. In: 2012 IEEE Conference on Computer Vision and Pattern Recognition, pp. 2288–2295, June 2012. https://doi.org/10.1109/CVPR.2012.6247939

10. Mairal, J., Bach, F., Ponce, J., Sapiro, G.: Online learning for matrix factorization and sparse coding. J. Mach. Learn. Res. **11**, 19–60 (2010). https://doi.org/10.1155/2013/259863

11. Mudunuri, S.P., Biswas, S.: Dictionary alignment for low-resolution and heterogeneous face recognition. In: 2017 IEEE Winter Conference on Applications of Computer Vision (WACV), pp. 1115–1123, March 2017. https://doi.org/10.1109/WACV.2017.129

12. Nikisins, O., Nasrollahi, K., Greitans, M., Moeslund, T.B.: RGB-D-T based face recognition. In: 2014 22nd International Conference on Pattern Recognition, pp. 1716–1721, August 2014. https://doi.org/10.1109/ICPR.2014.302

13. Ouyang, S., Hospedales, T., Song, Y.Z., Li, X., Loy, C.C., Wang, X.: A survey on heterogeneous face recognition: Sketch, infra-red, 3D and low-resolution. Image Vis. Comput. **56**, 28–48 (2016). https://doi.org/10.1016/j.imavis.2016.09.001

14. Qiu, Q., Chellappa, R.: Compositional dictionaries for domain adaptive face recognition. IEEE Trans. Image Process. **24**(12), 5152–5165 (2015). https://doi.org/10.1109/TIP.2015.2479456

15. Reale, C., Lee, H., Kwon, H.: Deep heterogeneous face recognition networks based on cross-modal distillation and an equitable distance metric. In: 2017 IEEE Conference on Computer Vision and Pattern Recognition Workshops (CVPRW), pp. 226–232, July 2017. https://doi.org/10.1109/CVPRW.2017.34

16. Reale, C., Nasrabadi, N.M., Chellappa, R.: Coupled dictionaries for thermal to visible face recognition. In: 2014 IEEE International Conference on Image Processing (ICIP), pp. 328–332, October 2014. https://doi.org/10.1109/ICIP.2014.7025065

17. Reyhanian, S., Arbabi, E.: Weighted vote fusion in prototype random subspace for thermal to visible face recognition. In: 2015 2nd International Conference on Pattern Recognition and Image Analysis (IPRIA), pp. 1–5, March 2015. https://doi.org/10.1109/PRIA.2015.7161647

18. Riggan, B.S., Reale, C., Nasrabadi, N.M.: Coupled auto-associative neural networks for heterogeneous face recognition. IEEE Access **3**, 1620–1632 (2015). https://doi.org/10.1109/ACCESS.2015.2479620

19. Sarfraz, M.S., Stiefelhagen, R.: Deep perceptual mapping for cross-modal face recognition. Int. J. Comput. Vis. **122**(3), 426–438 (2017). https://doi.org/10.1007/s11263-016-0933-2

20. Vu, T.H., Monga, V.: Learning a low-rank shared dictionary for object classification. In: 2016 IEEE International Conference on Image Processing (ICIP), pp. 4428–4432, September 2016. https://doi.org/10.1109/ICIP.2016.7533197

21. Vu, T.H., Monga, V.: Fast low-rank shared dictionary learning for image classification. IEEE Trans. Image Process. **26**(11), 5160–5175 (2017). https://doi.org/10.1109/TIP.2017.2729885

22. Wang, D., Kong, S.: A classification-oriented dictionary learning model: explicitly learning the particularity and commonality across categories. Pattern Recognit. **47**(2), 885–898 (2014). https://doi.org/10.1016/j.patcog.2013.08.004

23. Wei, C., Wang, Y.F.: Undersampled face recognition via robust auxiliary dictionary learning. IEEE Trans. Image Process. **24**(6), 1722–1734 (2015). https://doi.org/10.1109/TIP.2015.2409738

24. Xu, Y., Li, Z., Yang, J., Zhang, D.: A survey of dictionary learning algorithms for face recognition. IEEE Access **5**, 8502–8514 (2017). https://doi.org/10.1109/ACCESS.2017.2695239

25. Yang, M., Liu, W., Luo, W., Shen, L.: Analysis-synthesis dictionary learning for universality-particularity representation based classification. In: Thirtieth AAAI Conference on Artificial Intelligence (2016)

26. Yang, M., Zhang, L., Feng, X., Zhang, D.: Sparse representation based fisher discrimination dictionary learning for image classification. Int. J. Comput. Vis. **109**(3), 209–232 (2014). https://doi.org/10.1007/s11263-014-0722-8

27. Zhang, T., Wiliem, A., Yang, S., Lovell, B.: TV-GAN: generative adversarial network based thermal to visible face recognition. In: 2018 International Conference on Biometrics (ICB), pp. 174–181, February 2018. https://doi.org/10.1109/ICB2018.2018.00035

28. Zhu, F., Shao, L.: Weakly-supervised cross-domain dictionary learning for visual recognition. Int. J. Comput. Vis. **109**(1), 42–59 (2014). https://doi.org/10.1007/s11263-014-0703-y

Parking Slot Assignment for Overnight Electric Vehicle Charging Based on Network Flow Modeling

Junghoon Lee[✉] and Gyung-Leen Park

Department of Computer Science and Statistics, Jeju National University,
Jeju-si, Republic of Korea
{jhlee,glpark}@jejunu.ac.kr

Abstract. This paper designs a parking slot assignment scheme for electric vehicles (EVs) to support efficient overnight charging. Built upon a sophisticated computer algorithm, namely, network flow modeling, it takes EVs and parking areas as flow nodes. For an EV node, the link capacity to those parking place nodes within the walking distance limitation is set to 1. In addition, for a parking place node, the link capacity to the sink node is set to the number of parking slots facilitating EV charging. Moreover, the walking distance constraint is adjusted by the binary search mechanism to further reduce the worst-case distance from the parking place to the driver's home or lodge, taking advantage of affordable time complexity. The performance measurement result, obtained from a prototype implementation on the real-life parking area distribution, shows that the proposed scheme can satisfy 67.7% of parking requests for 10,000 EVs and average walking distance is around 70% of the maximum distance constraint.

Keywords: Electric vehicles · Overnight charging · Smart city · Parking slot assignment · Network flow model

1 Introduction

[1]Smart cities are capable of providing smart urban services, integrating computational intelligence coming from sophisticated computer algorithms, sensor technologies, big data analysis schemes, and the like [1]. Particularly, the penetration of diverse sensors into our everyday lives allows the physical world to be connected to the cyber world, where a computation entity grabs the current status of target objects and decides the corresponding reaction [2]. In addition, citizens can ubiquitously interact with each other and the computation entity via prevalent wireless communication channels by means of mobile devices, usually smart phones. Hence, it is possible for computation entities, as a form of

---

[1] Following are results of a study on the "Leaders INdustry-university Cooperation+" Project, supported by the Ministry of Education, (MOE), Republic of Korea.

© Springer Nature Switzerland AG 2019
R. Chamchong and K. W. Wong (Eds.): MIWAI 2019, LNAI 11909, pp. 181–190, 2019.
https://doi.org/10.1007/978-3-030-33709-4_16

high-end servers, to monitor the current status of target object, collects requests from many citizens, and activate computing tasks to provide an intelligent service capable of enhancing the quality of life.

While urban population growth demands many diverse smart city services, smart mobility is one of the most important components in modern cities, as a lot of citizens go to and back from their offices day by day [3]. Moreover, some heavily-crowded cities suffer from parking space insufficiency even in residential areas. Hence, parking slot assignment problems have been researched to optimize different objectives such as fairness or minimization of distance to the destination for each driver. However, they commonly have to take care of a massive number of vehicles and parking slots in city areas. Up to now, the cyber world has much faster time scale, compared with the physical world, so the computation time complexity generally does not impose a critical problem. On the contrary, modern services tend to combine diverse, sometimes heterogeneous datasets as many as possible, for example, weather conditions, event information, and the like. Hence, it is necessary to pursue more efficient computation strategies [4].

In the meantime, the transport system is now witnessing not so fast, but steady deployment of EVs (Electric Vehicles) with the appearance of smart grids in power networks [5]. They can reduce greenhouse gas emissions and air pollutions, fossil fuels. However, even though much improved than before, EVs have critical drawbacks in electricity charging, compared with their counterparts, namely, gasoline-powered vehicles. EV drivers want to avoid daytime charging, which makes drivers wait near the charger for tens of minutes mostly without doing anything. Instead, they prefer overnight charging with slow AC chargers, which are installed in public or private parking spaces. So, the parking slot assignment problem turns into the electricity charger assignment for EV drivers. The difference lies in that the charger assignment problem collects the charging requests by the end of the predefined deadline, runs the allocation procedure, and informs drivers of the result [6].

In this regard, this paper designs a charger or parking slot assignment scheme for EVs based on the well-known network flow model, specifically, the Ford-Fulkerson algorithm [7]. The proposed scheme takes the real-life parking slot distribution in Jeju City, Republic of Korea, which is actively deploying EVs under the citywide project called *Carbon Free Island*. A parking area has a different number of parking slots, while some are equipped with charging devices. An EV and a parking spot are connected when they are not far away from each other more than the given distance limit. Then, the flow model solver tries to enhance the number of EVs matched to parking slots, enforcing the distance between the request-issuing place and the slot to be less than the given limit value. Moreover, it exploits a binary search scheme to find an appropriate distance limit and thus reduce the worst-case walking distance for the given request set.

This paper is organized as follows: After outlining the main topic in Sect. 1, Sect. 2 briefly describes the background of this paper with the ongoing EV penetration effort in our city. Section 3 designs an allocation scheme focusing on how

Fig. 1. Parking area distribution

to model the assignment problem to the network flow model, Sect. 4 demonstrating the performance measurement results. Finally, Sect. 5 summarizes and concludes this paper with a brief introduction of future work.

2 Background

To begin with, the target city provides public parking spaces as shown in Fig. 1. Currently, there is no reservation or assignment mechanism for vehicles. However, with the wide deployment of EVs, it is necessary to build a facility capable of preventing unauthorized vehicles from occupying a slot. The area stretches about 10 km horizontally and 5 km vertically, respectively. This region is just a portion of the entire island and more than half of the population is concentrated in this area, its population reaching about 300,000. This city is carrying out an ambitious enterprise replacing all vehicles with EVs by 2030 [8]. Up to now, tens of thousands of EVs are on the road. The enterprise provides diverse charging facilities, be it fast or slow charging, especially on the public places. In the figure, 320 parking places are marked as circles, whose size coincides with their capacity. The largest one can accommodate 625 vehicles. During the night, EV drivers want park and charge their vehicles in the place as close to their homes as possible.

As an example of a parking slot assignment, [9] proposes a reservation-based mechanism, modeling the problem as multi-objective optimization. Those objectives taken into account include the walking distance to destination from parking place, the distance to parking place from request spot, and parking congestion impact. To cope with a large number of participating entities, this scheme takes a suboptimal approach on top of simulated annealing mechanisms. In addition,

EVs   Parking areas

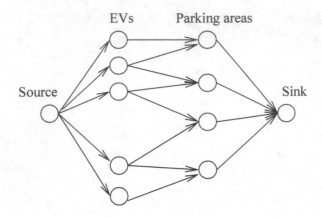

Source   Sink

Fig. 2. Network flow model

[10] proposes an assignment scheme addressing the fairness in the distance to the parking slot among each vehicle, somewhat sacrificing the system-wide efficiency, namely, the total sum of distances for the whole (vehicle-slot) pairs. A set of possibly conflicting constraints are selected in terms of the distance factor and the social benefits, and then a partial Langrangian is formulated by dualizing those constraints. Most approaches either need to assign slots on necessary basis or take a quite short allocation period. It is the difference from overnight parking and charging (Fig. 2).

3   Slot Assignment Design

Our system receives assignment requests by the end of a predefined deadline and runs an allocation procedure in the batch style. Non-admitted vehicles will either look for another place, possibly non-free, or give up overnight charging. An EV specifies the distance constraint to set the permissible range of distance between his or her home and the parking place. After all, there can be more than one candidate for each EV and the allocator matches as many EVs to charging places as possible. As we assume that it is not necessary to consider the precedence between the candidates, this problem can be solved by a network flow model. The flow model builds a directional graph, in which source and sink nodes are placed at the entry and end points of the network model. In addition, every link has a capacity to represent how much flow can be accommodated. It must be mentioned that some links are fully used, while others not.

In the slot assignment problem, the flow graph takes EVs and parking or charging places as graph nodes. After node definitions, edges are added from the source and respective EV nodes first. The link capacity for each link is just one as only an EV can be assigned to a single parking slot. If fluid can flow from the source to an EV node, the EV can be allocated. Next, for each parking area node, a link is added toward the sink. The capacity of a link coincides with the number of (charging-enabled) parking slots in the area. Hence, the capacity corresponds to the size of a circle in Fig. 1. The non-uniform link capacity makes this problem different from the bipartite matching procedure. Finally, an EV node can have a link to a parking area node as long as the distance between them is less than or equal to the distance limit. As the coordinate of each parking area is given as an open dataset, we can calculate the Euclidean distance from a vehicle (or home) to the parking area via simple mathematical calculation. With this modeling, a network flow solver, such as the Ford-Fulkerson algorithm, can find the maximum amount of flow from the source to the sink.

The distance constraint may be different EV by EV or can be given as a flat value over the whole system. When the constraint is loose, the number of links increases and the assignment procedure can benefit from wider options. Here, the time complexity is dependent on the number of links and the maximum permissible flow. Even in the case of hundreds of thousands of links, the computation time doesn't matter. After an assignment procedure, we can obtain the average and maximum distance among all (EV, slot) pairs. Hence, when the distance constraint is given as a system parameter, those distances can be further improved by tuning the distance constraint and re-executing the network flow solver.

The number of matched pairs with the initially given distance constraint is stored as a basis value. The adjustment procedure tries different values to reduce the distance constraint, according to the binary search with the lower and upper bounds set to 0 and the basis value. Each time the solver completes, the number of matched pairs is compared with the basis. If the newly calculated number of pairs is smaller, the lower bound will be increased by the half of the current search interval. On the contrary, if equal, the upper bound is cut down by the same amount. The size of the search interval is halved in each step. If there is no possibility to further reduce the constraint, both lower and upper bounds reach the basis value. This step lets a closer pair be selected when multiple pairs can be included in the final set, instead of randomly selecting.

4   Performance Analysis

This section measures the performance of the proposed assignment scheme using the real-life parking area distribution, in which 320 parking areas have 14,653 slots in total. The network flow solver is implemented as a Java program. The execution time of the solver is less than 1 second for tens of thousands of nodes on average performance PCs. As it is not possible to accurately estimate the parking space demand from the EV side, the experiment assumes that assignment requests are generated randomly over the residential area in the target city. In the experiment, main performance parameters are the length of distance constraints and the number of EVs while the performance metrics include the number of matched pairs as well as system-wide average and maximum distance between (EV, slot) pairs, respectively.

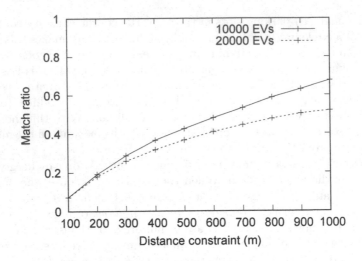

Fig. 3. EV match ratio vs. distance constraint range

First, Fig. 3 plots the effect of the distance constraint to the matched ratio. The experiment tracks two cases of 10,000 EVs and 20,000 EVs. As the total number of parking slots is fixed to 14,653, not all EVs can be assigned in the latter case. In the case of 10,000 EVs, where more parking capacity is available, up to 67.7% of requests can be admitted when the distance constraint is 1 km. Actually, in this distance, another personal transport method such as bicycles or kickboards may be necessary. On the contrary, in the case of 20,000 EVs, the degree of increment is quite limited in spite of the enlarged distance constraint, the EV match ratio reaching 51.9% of EV admission. From the viewpoint of parking areas, the slot occupancy reaches 70.9% in the 20,000 EV case with 1 km constraint, as shown in Fig. 4. Slot occupancy gets much increased with more EVs.

The next experiment measures the average and maximum distance in the set of matched pairs with the same parameter setting with the previous experiment. Figures 5 and 6 show that there is no difference in both metrics for 10,000 and 20,000 EV cases. The almost same average distance, which is independent of the number of EVs, shows that the network flow solver commonly picks those links having the similar capacity even when there are multiple choices. According to Fig. 5, we can say that a pair is more likely selected when it just meets the constraint, as the solver always try to maximize the number of matched pairs. Additionally, the average-maximum ratio ranges from 0.68 to 0.76 in the 10,000 EV case and from 0.66 to 0.78 in the 20,000 EV case, respectively. Even with more EVs, the ratio is not affected and the difference is negligible.

Now, let the performance parameter be changed. Figure 7 plots the effect of the number of EVs while the distance constraint is fixed to 500 m. In this constraint range, EV drivers can generally go to their homes or other destinations on foot. Figure 7 shows that the match ratio decreases according to the increase

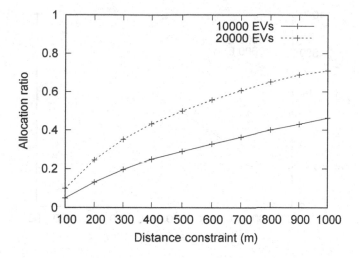

Fig. 4. Slot allocation ratio vs. distance constraint

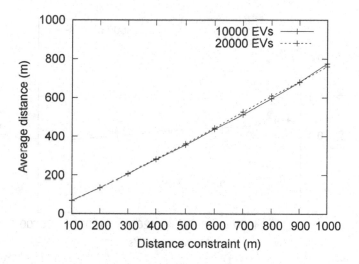

Fig. 5. Average distance vs. distance constraint

in the number of EVs, in spite of the increase in the number of matched pairs. There may exist a saturation point which corresponds to the total number of parking slots or below. However, the degradation is quite linear. According to the figure, the match ratio begins from 0.48 and decreases gradually, reaching 0.36 with 20,000 EVs. Finally, Fig. 8 plots the average distance of (EV, slot) pairs in the matched set. It is not necessary to plot the maximum distance as it

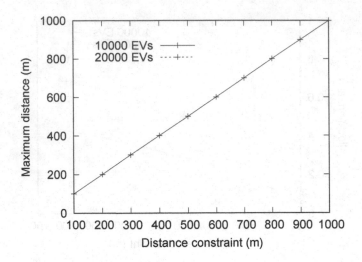

Fig. 6. Maximum distance vs. distance constraint

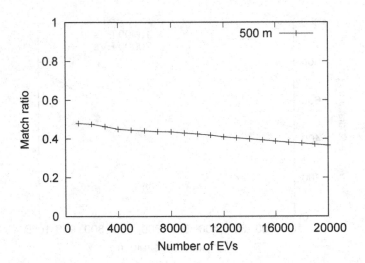

Fig. 7. Match ratio vs. number of EVs

will be almost same as the distance constraint, as explained previously. Figure 8 again shows that the average distance is not influenced by the number of EVs, while the average distance ranges from 343 to 361 m for the distance constraint of 500 m. The average tends to get improved according to the increase in the number of EVs, as more EVs are matched.

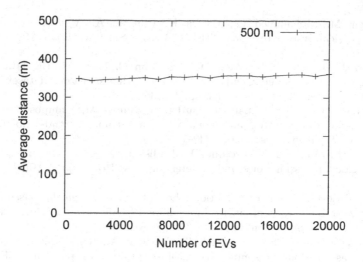

Fig. 8. Average distance vs. number of EVs

5 Conclusion

In this paper, we have designed a parking slot assignment scheme for EVs to efficiently supply energy via overnight charging. The charger-facilitated slot assignment is formed as a network flow model consisting of EV and parking area nodes as well as inter-party links. A link between an EV node and a parking area node is added according to the distance between them. The performance measurement result, obtained by the experiment exploiting a real-life parking space distribution, showed that the proposed scheme can satisfy 67.7% of parking requests for 10,000 EVs and average walking distance is around 70% of the maximum distance constraint. It is an example of applying an intelligent computer algorithm to the smart city environment. Even though the number of entities involved in the computing process increases, the computation capacity and intelligence can sufficiently cope with this problem. However, it is necessary to support tight control to implement a computer-created assignment policy, for example, the preventing unauthorized EVs from occupying charging slot and the like.

As future work, we are planning to design a virtual private power network, which supports personalized energy system mainly consisting of renewable energy generators, electricity consumers, and batteries, leasing the part of public networks. It will also take the network flow model and can easily return rewards to individuals for optimized energy consumption, renewable energy generation, and the like.

References

1. Colistra, J.: The evolving architecture of smart cities. In: Proceedings of 4th IEEE International Smart Cities Conferences (2018)

2. Cintuglu, M., Mohammed, O., Akkaya, K., Uluagac, A.: A survey on smart grid cyber-physical system testbeds. IEEE Commun. Surv. Tutorials **19**(1), 446–464 (2017)
3. Zenkert, J., Drnerhofer, M., Weber, C., Ngoukam, C., Fathi, M.: Big data analytics in smart mobility: modeling and analysis of the Aarhus smart city dataset. In: IEEE Industrial Cyber-Physical Systems, pp. 363–368 (2018)
4. Torre-Bastida, A., Ser, J., Lana, I., Ilardia, M., Bilbao, M., Campos-Cordobes, S.: Big data for transportation and mobility: recent advances, trends and challenges. IET Intel. Transport Syst. **12**(8), 742–755 (2018)
5. Masera, M., Bompard, E., Profumo, F., Hadjsaid, N.: Smart (electricity) grid for smart cities: assessing roles and societal impacts. Proc. IEEE **106**(4), 613–625 (2016)
6. Lee J., Park, G.: Charger reservation scheme for electric vehicles based on stable marriage problem adaptation. In: International Conference on Information Technology: IoT and Smart City, pp. 316–319 (2018)
7. Benda, D., Chu, X., Sun, S., Quek, T., Buchley, A.: Renewable energy sharing among base stations as a min-cost-max-flow optimization problem. IEEE Trans. Green Commun. Netw. **3**(1), 67–78 (2019)
8. Lee, J., Park, G., Han, Y., Yoo, S.: Big data analysis for an electric vehicle charging infrastructure using open data and software. In: Proceedings of ACM eEnergy, pp. 252–253 (2017)
9. Mejri, N., Ayari, M., Langar, R., Saidane, L.: Reservation-based multi-objective smart parking approach for smart cities. In: Proceedings of 2nd IEEE International Smart Cities Conferences (2016)
10. Afonsetti, E., Weeraddana, P., Fischione, C.: Min-max fair car-parking slot assignment. In: Proceedings of IEEE 16th International Symposium on a World of Wireless, Mobile and Multimedia Networks (2015)

AIBA: An AI Model for Behavior Arbitration in Autonomous Driving

Bogdan Trăsnea[1,2]([✉]), Claudiu Pozna[1,3], and Sorin M. Grigorescu[1,2]

[1] Robotics, Vision and Control Lab, Transilvania University of Brasov,
Braşov, Romania
{bogdan.trasnea,cp,s.grigorescu}@unitbv.ro
[2] Department of Artificial Intelligence, Elektrobit Automotive, Timişoara, Romania
[3] Department of Informatics, Széchenyi István University of Győr, Győr, Hungary

Abstract. Driving in dynamically changing traffic is a highly challenging task for autonomous vehicles, especially in crowded urban roadways. The Artificial Intelligence (AI) system of a driverless car must be able to arbitrate between different driving strategies in order to properly plan the car's path, based on an understandable traffic scene model. In this paper, an AI behavior arbitration algorithm for Autonomous Driving (AD) is proposed. The method, coined AIBA (AI Behavior Arbitration), has been developed in two stages: (i) human driving scene description and understanding and (ii) formal modelling. The description of the scene is achieved by mimicking a human cognition model, while the modelling part is based on a formal representation which approximates the human driver understanding process. The advantage of the formal representation is that the functional safety of the system can be analytically inferred. The performance of the algorithm has been evaluated in Virtual Test Drive (VTD), a comprehensive traffic simulator, and in GridSim, a vehicle kinematics engine for prototypes.

Keywords: Behavior arbitration · Context understanding · Autonomous driving · Self-driving vehicles · Artificial intelligence

1 Introduction and Related Work

Autonomous Vehicles (AVs) are robotic systems that can guide themselves without human operators. Such vehicles are equipped with *Artificial Intelligence* (AI) components and are expected to change dramatically the future of mobility, bringing a variety of benefits into everyday life, such as making driving easier, improving the capacity of road networks and reducing vehicle-related accidents.

Most likely, due to the lack of safety guarantees and legislation, as well as the missing scalability of *Autonomous Driving* (AD) systems, fully autonomous vehicles will not travel the streets in the near future. Nevertheless, over the past years, the progress achieved in the area of AI, as well as the commercial availability of Advanced Driver Assistance Systems (ADAS), has brought us closer to the goal of full driving autonomy [1,2].

© Springer Nature Switzerland AG 2019
R. Chamchong and K. W. Wong (Eds.): MIWAI 2019, LNAI 11909, pp. 191–203, 2019.
https://doi.org/10.1007/978-3-030-33709-4_17

The Society of Automotive Engineers (SAE) has defined standard J3016 for different autonomy levels [3], which splits the concept of autonomous driving into 5 levels of automation. Levels 1 and 2 are represented by systems where the human driver is required to monitor the driving scene, whereas levels 3, 4 and 5, with 5 being fully autonomous, consider that the automated driving components are monitoring the environment.

An AV must be able to sense its own surroundings and form an environment model consisting of moving and stationary objects [4]. Afterwards it uses this information in order to learn long term driving strategies. These driving policies govern the vehicle's motion [5] and automatically output control signals for steering wheel, throttle and brake. At the highest level, the vehicle's decision-making system has to select an optimal route from the current position to the destination [6].

The main reason behind the human ability to drive cars is our capability to understand the driving environment, or driving context. In the followings, we will refer to the driving context as the scene and we will define it as linked patterns of objects. In this paper, we introduce AIBA (*AI Behavior Arbitration*), an algorithm designed to arbitrate between different driving strategies, or vehicle behaviors, based on AIBA's understanding of the relations between the scene objects.

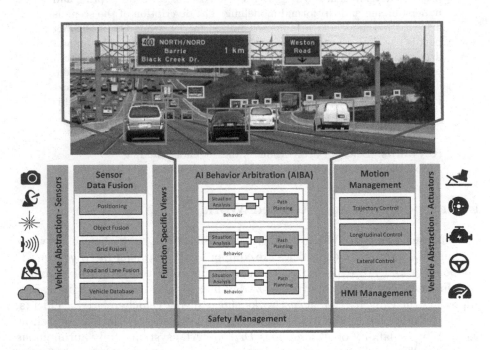

Fig. 1. Behavior Arbitration using AIBA in the EB robinos autonomous driving framework

A driving scene consists of objects such as lanes, sidewalks, cars, pedestrians, bicycles, traffic signs, etc., all of them being connected to each other in a particular way (e.g. a traffic sign displays information for a driver). An example of such a driving scene, processed by AIBA within EB robinos, is depicted in Fig. 1. EB robinos is a functional software architecture from Elektrobit Automotive GmbH, with open interfaces and software modules that manages the complexity of autonomous driving. The EB robinos reference architecture integrates the components following the *sense, plan, act* decomposition paradigm. Moreover, it also makes use of AI technology within its software modules in order to cope with the highly unstructured real-world driving environment.

Driving behavior prediction is a key part of an AV's decision-making process. The task of identifying the future behavior of the scene's objects is not a trivial one, since this type of information cannot be directly measured or communicated, being considered latent information [7]. To be able to perform behavior prediction for the surrounding vehicles, an AV can use mathematical models which consider the variation in the objects' movement and describe the driving scenario from the view point of the ego-car.

Such kind of models use several types of information, i.e. vehicle kinematics, the relationship between the ego-car and the surrounding entities, the interactions with other vehicles and a-priori knowledge. Vehicle kinematics and the relations with road entities were considered by almost all existing studies [8]. For example, Lefevre et al. used the Time to Line Crossing (TTLC) to predict whether the vehicle will depart from the current lane or not [9].

Usually, most models for behavior arbitration are tailored for one specific scenario. Nevertheless, AVs must drive through dynamically changing environments in which a diversity of scenarios occur over time [10]. Multiple scenario-specific models activate a corresponding model according to the characteristics of the scenario. In [11], the authors of this paper have proposed a three level Milcon type architecture. The second level, called the tactical level, includes a behavior collection and a manager program which trigger the appropriate AV behavior.

In this paper, we introduce AIBA, a system for behavior arbitration in AVs, which constructs a description and understanding model of the driving scene. Our idea is to model the human driver (HDr) understanding process of the driving scene, in order to achieve an optimal behavior arbitration solution. We describe the mechanism of the HDr thinking and transpose it to an approximate model.

The rest of the paper is organized as follows. The concept for describing the driving scene is given in Sect. 2, followed by the understanding and modelling presented in Sect. 3. The experiments showcasing AIBA are detailed in Sect. 4. Finally, conclusions are stated in Sect. 5.

2 Driving Scene Description

The driving scene description is given from a human driver's perspective, and it formulates properties derived from the definitions of classes, subclasses and objects which represent the core of an abstraction model, based on the authors' previous work [12]. The main idea behind AIBA is to model, or formalize, the HDr understanding process and afterwards transform it into a formal model for behavior arbitration in AV.

2.1   Driving Scene Analysis

A human driver is able to perceive the scene's objects and observe them. This means that the HDr identifies the concepts and the different properties of the objects. In the end the driver can describe the scene in Natural Language (NL), as for the traffic example in Fig. 1: "the car is near exit E1; the main road continues straight and will reach the location L1, L2, L3; the traffic has a medium density and takes place under normal circumstances; the car is in the 3rd lane in safe vicinity to other cars". Apparently, this message eludes a lot of information, but if it is shared with other human drivers, it proves to be a piece of very important and comprehensive knowledge.

The different scene objects are linked between them, as illustrated in the scene network from Fig. 2, built for the traffic scene shown in Fig. 1. Here, intuition assumes that the most important links are those established with the front car, or the main road (bold red lines in the image), while the less important links are those between the ego-car and the buildings (thin red lines in the image). In Fig. 2, each class of objects is linked with a rectangle. The symbols used are: AC for the ego-vehicle autonomous car, Bld for the buildings (i.e. side barriers), $C1$, $C2, \ldots, Cn$ for the other traffic cars, M, R, FR for lane identification, and RS for the road traffic signs.

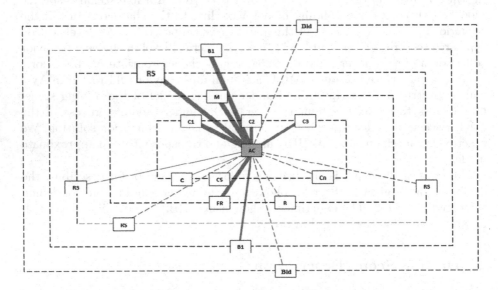

Fig. 2. Driving scene network representing links between different objects, as imagined by a human driver. Bold red lines suggest a higher link importance level, whereas thin red lines point to a lower importance level. (Color figure online)

The links definition is a first step in the knowledge process, which means it establishes the subjects of interest and also the importance level of each subject. The HDr scene understanding synthesis contains the following steps: identifying the link between the objects, allocating models to each link, running the models and afterwards finding a strategy to act. In fact, the HDr creates an implicit system and simulates it.

Two types of links can be observed: internal links, considered to be the set of links between the observer (the HDr) and all the observed objects, and external links, represented by the set of links between the objects present in the scene. From the HDr point of view, the links have different meanings and importance, or significance. Specifically, a human driver knows the traffic rules and how to get to his destination. These rules will determine traffic priorities, thus making the HDr to attach a greater importance to those links which are more important to his strategy.

In the next step, a description is established for each link. During the driving, the HDr adapts, or refines, the mentioned description by observation. The driver, by using the importance of the links, simulates a scene description. If we analyze the aim of driving scene understanding, we will observe that its origins are the stability in time and space. More precisely, the human driver has a driving task which can be accomplished if the possibility of locomotion is preserved in the current position and during a specific time. Intuitively, the stability is related to the objects in the scene and can be reached by understanding the scene. This understanding does not solve the driving problem, but offers the information upon which the human driver decides to act on a certain behavior.

2.2 Description Proposal

Several concepts can be proposed at this point of analysis:

- The understanding is related with the concept of intelligence which is defined by Piaget as "the human ability to obtain stability in time and space" [13]. This means that the links and the models (behaviors) are correlated to the HDr intelligence.
- The driving task has multiple attributes: determine a trajectory from start to end, ensure the comfort, minimize the energy consumption etc., all which are time dependent. The task trajectory must be accomplished in safe conditions and should be included in the stability concept.
- The driving scene is a complex environment, many entities being linked. In order to handle this complexity, the HDr establishes a network representing a holistic interpretation and priority interactions for each link of the network
- Anticipating, the AD system is equipped with a sensor network, giving highly qualitative information about the scene.

Resuming, the previous analysis of the driving scene understanding can be described like a process which consists of the following steps:

1. Perceiving the objects (cars, traffic signs, lanes, pedestrians, etc.) and obtaining the object properties (the pedestrian intention is to cross the road).
2. Defining the links (the scene network) between the objects and their importance (i.e. the car in front is important, the pedestrian which intends to cross the road is very important, etc.).
3. Adding models to the mentioned links.
4. Simulating the model of the most significant links, and proposing driving behaviors which will be verified with the other significant links until the appropriate behavior (which allows the driving task) is found.

3   Driving Scene Modelling

Entities like objects, links, or networks, which have been introduced in the previous section, have correspondences in the modelling process. Our intention is to approximate the HD understanding process through a formal representation. More precisely, this assignment mimics an input/output process: using a perceived scene, the AIBA model must output a description which offers all the information needed within the AD system to arbitrate the driving behavior.

The block diagram representation of AIBA is illustrated in Fig. 3, where the information flow emulates the human driver scene understanding phenomenon.

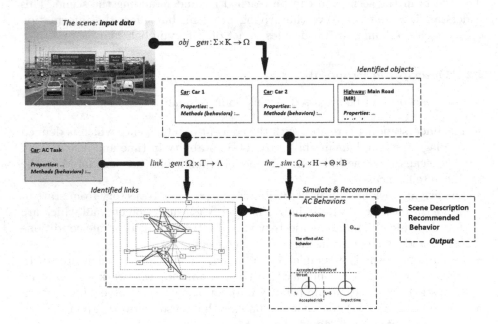

Fig. 3. The block diagram of AIBA, overall picture.

Within AIBA, the first action is to transform the scene in to a collection of objects. This operation is accomplished by the following generative function:

$$obj\_gen : \Sigma \times K \to \Omega \tag{1}$$

where Σ is the perceived scene; K is the set of known object classes and Ω is the set of objects.

Having an initial collection of classes, the generator function from Eq. 1 transforms the scene entities into a collection of objects. The following object classes are taken into consideration: traffic participants, pedestrians and buildings. The set of classes are a priori set within AIBA.

The complexity of this process, even for the set of road classes, is seen in many types of roads which exist around the world. In order to reduce the scene's complexity, we split them in two major classes, static objects (lanes, traffic signs, buildings, etc.) and dynamic objects (cars, pedestrians, etc.).

Two kinds of definitions are here significant: the generic definition where the proximity and the properties are mentioned, and the extensive definition where the definition of the object (or a picture of it) is indicted or shown (similar to Fig. 1). This observation enables us to associate image recognition methods (which correspond to the extensive definition) with a generic definition collection.

Equation 1 can be generalized when measurements (speed of the cars, distance between cars, size of traffic signs etc.) are associated with the scene description:

$$obj\_gen : \Sigma \times K \times M \to \Omega, \tag{2}$$

where M is the set of measurements. The generated objects Ω contain a structure of class specific properties and methods. The methods reflect the possible interactions of the objects with the ego-car.

The second step in AIBA's workflow defines the links between a IIDr and an object, as well as between the objects themselves, respectively:

$$link\_gen : \Omega \times T \to \Lambda, \tag{3}$$

where T is the set of task trajectory properties and Λ is the set of links' significance.

Because the links are computed in term of stability around a particular point of the task trajectory, the significance is correlated with specific threats. Driving a car is subject to implicit negotiation based on traffic rules. The most important links are those related to agents which, according to these rules, have priority. Equation 3 can be imagined as an expert system which will analyse all these links from the mentioned point of view, while outputting different marks:

$$obj\_gen : \Sigma \times K \times M \times T \to \Omega \times \Lambda. \tag{4}$$

Each recognized object provides methods which refer to the ego-car - object interactions. Information about possible threats on the task stability can be obtained by simulating the behaviors, and also how to select the appropriate driving behavior for avoiding these threats:

$$thr\_sim : \Omega_S \times H \rightarrow \Theta \times B, \tag{5}$$

$$\Omega_S = \{O_i | O_i \in \Omega; S_{O_i,O_j} \geq S_{min}\} \tag{6}$$

$$H = [t_c \quad t_c + \delta] \tag{7}$$

where Ω_S is the set of important objects O_i, which are linked with other objects O_j with a significance $S_{O_i,O_j} \in \Lambda$, greater than a minimum (a priori imposed) significance S_{min}. H is the time horizon, t_c is the current time, δ is the simulation time, Θ is the set of the threat levels and B is the set of recommended behaviors for the ego-car.

thr_sim will simulate, for each important object O_i, a collection of behaviors $O_i\_M_{i,k}(P, H)$ and a predicted threat level $\Theta_{i,k}$:

$$O_i\_M_{i,k}(P, H) = \begin{bmatrix} \Theta_{i,k} \\ \beta_{i,k} \end{bmatrix} \tag{8}$$

where P is the set of object properties and $\beta_{i,k}$ is the behavior of the ego-car which will eliminate the threat. The mentioned models can be analytical functions [2], fuzzy engines [14], or even neural networks [15].

In order to solve all links' threats, several strategies can be chosen. If we adopt the HDr understanding description from the previous section, the behavior which solves the maximum threat is simulated for the other threats, obtaining the optimal behavior of the ego-car $\Theta_{max} = \Theta_{i^*,k^*}$:

$$(i^*, k^*) = \arg \max_{i,k} \Theta_{i,k} \tag{9}$$

where $\Theta_{i,k}$ are the threat levels.

The last step in AIBA's modelling system is the transformation of the optimal behavior Θ_{max} into a natural language explanation:

$$dsc : \Lambda \times B \rightarrow \Delta \tag{10}$$

$$\Delta = \Omega_S \times E \tag{11}$$

$$E = e_1 \times e_2 \times ... \times e_{n_S} \tag{12}$$

where Δ is the set of descriptions, e_i is the explanation of the significance associated to an object and n_S is the number of significant objects.

4   Experiments

Our experiments were conducted in our own autonomous vehicles prototyping simulator GridSim [17] and in Virtual Test Drive(VTD).

GridSim is a two dimensional birds' eye view autonomous driving simulator engine, which uses a car-like robot architecture to generate occupancy grids from simulated sensors. It allows for multiple scenarios to be easily represented and loaded into the simulator as backgrounds, while the kinematic engine itself is based on the single-track kinematic model.

In such a model, a no-slip assumption for the wheels on the driving surface is considered. The vehicle obeys the "non-holonomic" assumption, expressed as a differential constraint on the motion of the car. The non-holonomic constraint restricts the vehicle from making lateral displacements, without simultaneously moving forward.

Occupancy grids are often used for environment perception and navigation, applications which require techniques for data fusion and obstacles avoidance. We used such a representation in the previous work for driving context classification [18]. We assume that the driving area should coincide with free space, while non-drivable areas may be represented by other traffic participants (dynamic obstacles), road boundaries, buildings, or other static obstacles. The virtual traffic participants inside GridSim are generated by sampling their trajectory from a uniform distribution which describes their steering angle's rate of change, as well as their longitudinal velocity. GridSim determines the sensor's freespace and occupied areas, where the participants are considered as obstacles. This representation shows free-space and occupied areas in a bird's eye perspective. An example of such a representation can be seen in Fig. 4.

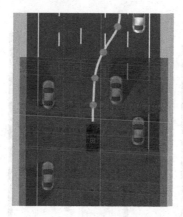

Fig. 4. GridSim example of a simulation scenario. The green area is a simulated occupancy grid, used to enhance the measurement of the properties of objects in the scene. (Color figure online)

VTD is a complete tool-chain for driving simulation applications. The toolkit is used for the creation, configuration, presentation and evaluation of virtual environments in the scope of based simulations. It is used for the development of ADAS and automated driving systems as well as the core for training simulators. It covers the full range from the generation of 3D content to the simulation of complex traffic scenarios and, finally, to the simulation of either simplified or physically driven sensors.

Fig. 5. VTD - Virtual Test Drive example of a simulation scenario.

For better understandability, we explain the behavior arbitration steps on a proposed simulation scenario. The sketch of the such a scenario can be found in Fig. 6. The car is driving in the middle lane, behind a car in the same lane. A car from the left side signals its intention to change the lane to the right, in front of the ego-vehicle.

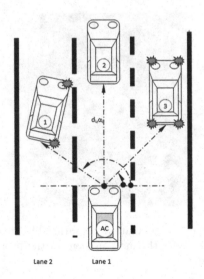

Fig. 6. Sketch of the simulation scenario

In the first step the ego-vehicle identifies the three objects from the scene and measures the properties, also computing the possible impact times:

- Car 1: in lane 3 (left, near), at distance $d1$, with speed $v1$, changing the lane

- Car 2: in lane 2 (in front, far), at distance $d2$, with speed $v2$, following the 2nd lane
- Car 3: in lane 1 (right, near), at distance $d3$, with speed $v3$, intending to stop

Table 1. Table with resulted impact times. The highlighted values are the ones that pose threats to the ego-vehicle.

Car	LaneFollow	LaneChangeRight	LaneChangeLeft	Stop
	Impact Time			
1	0,2	0,59	0,01	0,2
	∞	20	∞	∞
2	0,6	0,1	0,1	0,2
	25	∞	∞	10
3	0,2	0,2	0,1	0,5
	∞	∞	30	∞

In the second step the ego-vehicle identifies links level:

- Car 1 (left, near) $\lambda_1 = 0.3$
- Car 2 (in front, far) $\lambda_2 = 0.6$
- Car 3 (right, near) $\lambda_3 = 0.1$

In the third step the threat time is computed, and compared to the maximum accepted level:

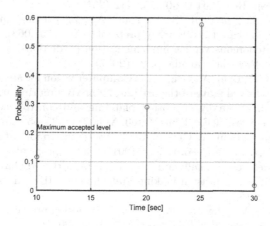

Fig. 7. Maximum accepted level of all threats, on a time axis.

In the fourth and final step, the behavior of the ego-vehicle which solves the most probable threat is reducing speed. This is also simulated for the other threats which are above the maximum accepted level, and since it also solves the lane change to the right of the first car, it is chosen as the optimal behavior of the ego-car.

5 Conclusion

In summary, this paper proposes a novel approach to address the task of behavior arbitration. The method, coined AIBA, aims at formalizing and learning human-like driving behaviors from the description, understanding and modelling of a traffic scenario. The main advantage of the formal representation is that the functional safety of the automotive system can be analytically inferred, as opposed to a neural network black-box, for example.

As future work, we would like to extend the proposed method to more granular driving scenarios, such as driving in an indoor parking lot, or driving on uncharted roads. Additionally, we would like to add more objects in the scene, and to test the deployment on an embedded device.

Acknowledgments. We hereby acknowledge Elektrobit Automotive, Széchenyi István University, and the TAMOP - 4.2.2.C-11/1/KONV-2012-0012 Project "Smarter Transport - IT for co-operative transport system" for providing the infrastructure and for support during research.

References

1. Katrakazas, C., Quddus, M., Chen, W.H., Deka, L.: Real-time motion planning methods for autonomous on-road driving: state-of-the-art and future research directions. Transp. Res. Part C **60**, 416–442 (2015)
2. Shalev-Shwartz, S., Shammah, S., Shashua, A.: On a formal model of safe and scalable self-driving cars (2017). arXiv preprint arXiv:1708.06374
3. Litman, T.: Autonomous vehicle implementation predictions. Victoria Transport Policy Institute, Victoria, Canada, p. 28 (2017)
4. Janai, J., Güney, F., Behl, A., Geiger, A.: Computer vision for autonomous vehicles: problems, datasets and state-of-the-art (2017). arXiv preprint arXiv:1704.05519
5. Paden, B., et al.: A survey of motion planning and control techniques for self-driving urban vehicles. IEEE Trans. Intell. Veh. **1**(1), 33–55 (2016)
6. Janai, J., Guney, F., Behl, A., Geiger, A.: Computer vision for autonomous vehicles: Problems, datasets and state-of-the-art (2017). arXiv preprint arXiv:1704.05519
7. Gindele, T., Brechtel, S., Dillmann, R.: Learning driver behavior models from traffic observations for decision making and planning. IEEE Intell. Transp. Syst. Mag. **7**, 69–79 (2015)
8. Lefèvre, S., Dizan, V., Christian, L.: A survey on motion prediction and risk assessment for intelligent vehicles. Robomech J. **1**, 1–14 (2014)
9. Lefevre, S., Gao, Y., Vasquez, D., Tseng, H.E., Bajcsy, R., Borrelli, F.: Lane keeping assistance with learning-based driver model and model predictive control. In: Proceedings of the 12th ISAVC, Tokyo, Japan (2014)

10. Geng, X., Liang, H., Yu, B., Zhao, P., He, L., Huang, R.: A scenario-adaptive driving behavior prediction approach to urban autonomous driving. Appl. Sci. **7**(4), 426 (2017)
11. Pozna, C., Troester, F.: Human behavior model based control program for ACC mobile robot. Acta Polytechnica Hungarica **3**(3), 59–70 (2006)
12. Pozna, C., Precup, R.E., Minculete, N., Antonya, C., Dragos, C.A.: Properties of classes, subclasses and objects in an abstraction model. In: 19th International Workshop on Robotics in Alpe-Adria-Danube Region, pp. 291–296 (2010)
13. Piaget, J.: The Psychology of Intelligence. Taylor & Francis, London (2005). ISBN-10: 0415254019
14. Wang, X., Fu, M., Ma, H., Yang, Y.: Lateral control of autonomous vehicles based on fuzzy logic. Control Eng. Pract. **34**, 1–17 (2016)
15. Bojarski, M., et al.: End to end learning for self-driving cars. CoRR, vol. abs/1604.07316 (2016)
16. Grigorescu, S.: Generative one-shot learning (GOL): a semi-parametric approach for one-shot learning in autonomous vision. In: International Conference on Robotics and Automation ICRA 2018, 21–25 May 2018, Brisbane, Australia (2018)
17. Trasnea, B., Marina, L.A., Vasilcoi, A., Pozna, C., Grigorescu, S.: GridSim: a vehicle kinematics engine for deep neuroevolutionary control in autonomous driving. In: 3rd IEEE International Conference on Robotic Computing (IRC), pp. 443–444 (2019)
18. Marina, L.A., Trasnea, B., Cocias, T., Vasilcoi, A., Moldoveanu, F., Grigorescu, S.: Deep grid net (DGN): a deep learning system for real-time driving context understanding. In: 3rd IEEE International Conference on Robotic Computing (IRC), pp. 399–402 (2019)

A Hierarchical Classification Method Used to Classify Livestock Behaviour from Sensor Data

Hari Suparwito[1(✉)], Kok Wai Wong[1], Hong Xie[1], Shri Rai[1], and Dean Thomas[2]

[1] Murdoch University, Perth, WA, Australia
shirsj@jesuits.net,
{K.Wong,H.Xie,S.Rai}@murdoch.edu.au
[2] CSIRO Floreat, Perth, WA, Australia
Dean.Thomas@csiro.au

Abstract. One of the fundamental tasks in the management of livestock is to understand their behaviour and use this information to increase livestock productivity and welfare. Developing new and improved methods to classify livestock behaviour based on their daily activities can greatly improve livestock management. In this paper, we propose the use of a hierarchical machine learning method to classify livestock behaviours. We first classify the livestock behaviours into two main behavioural categories. Each of the two categories is then broken down at the next level into more specific behavioural categories. We have tested the proposed methodology using two commonly used classifiers, Random Forest, Support Vector Machine and a newer approach involving Deep Belief Networks. Our results show that the proposed hierarchical classification technique works better than the conventional approach. The experimental studies also show that Deep Belief Networks perform better than the Random Forest and Support Vector Machine for most cases.

Keywords: Machine learning · Hierarchical classification · Livestock behaviour · Sensor data

1 Introduction

The use of computer and wireless sensors has shown potential in farm management systems [1]. This approach offers advantages in livestock businesses of 24/7 monitoring of animals without regular human observation. Monitoring of behaviour can assist with understanding how livestock meet production requirements in farming systems [2], and improves livestock welfare [3]. Behaviour monitoring can help identify ways to minimise the environmental footprint of the business [4], and the automation of some manual tasks can lead to more efficient farm management [5].

Knowledge of livestock behaviours provides important information about the livestock themselves and their environment. There is active research to classify livestock activities such as grazing, walking and resting based on wearable electronic sensors [6–8]. Recently, data from monitoring livestock behaviours has become more

© Springer Nature Switzerland AG 2019
R. Chamchong and K. W. Wong (Eds.): MIWAI 2019, LNAI 11909, pp. 204–215, 2019.
https://doi.org/10.1007/978-3-030-33709-4_18

diverse and complex with developments in on-animal wearables. Sensors and digital technologies have enabled us to monitor livestock behaviours without the need for human involvement [8–14]. Such technologies are capable of producing massive amount of data, which can be analysed using machine learning techniques to uncover hidden patterns. Knowledge of these patterns allows us to develop predictive modelling and intelligent decision support information based on empirical data.

Current research on the classification of livestock behaviours using machine learning classifies livestock behaviours on a one level behaviour classification [9, 11, 15]. The one level behaviour classification means all animal behaviour such as grazing, walking, resting, and drinking are directly observed and analysed at the same level. Unlike previous studies, we propose a hierarchical behaviour classification method. Instead of classifying all the animal behaviours at one level, we are performing classification in two levels. At the first level, behaviours are divided into Active and Inactive. In the second level, the Active behaviour from level one is further classified into two behaviours: grazing and walking. Similarly, the Inactive behaviour from level one is considered as "camping" and is sub-divided into two level two behaviours: Camping Active and Camping Inactive. Camping Active and Camping Inactive depict the livestock are moving in a small area or within a short distance. These also include stopping (standing or lying) in a specific location. The difference between the two inactive behaviours is based amount of movement within a small area where animals are observed to be resting, with the Camping Inactive behaviour having the lower amount of movement or no movement.

Previous studies on livestock behaviours [6, 8, 10] showed that, of all the methods tested, Random Forest (RF), and Support Vector Machine (SVM), were the best [16, 17]. Beside these two methods, we introduce the use of a Deep Learning method and compare the performance of these three methods in this paper [18]. Given that deep learning has emerged as another machine learning technique, this paper investigated the use of deep learning in the proposed hierarchical classification method. The main reason for using the Deep Learning method is its ability to discover features amenable for prediction by recursively applying a simple but nonlinear transformation to the data [19, 20]. We applied the three machine learning methods to analyse the sensor data for classifying livestock behaviours. However, the difference is that we compared the performance of the three machine learning methods for hierarchical classification of livestock behaviours while previous work in this area were all one level classifications.

We also examined whether the hierarchical classification approach would produce more accurate classifications compared to those using one level classification. It is expected that this study will contribute to a new method for the determination of livestock activities based on their behaviour classification by analysing the dataset obtained from animal collars containing GPS and accelerometer sensors. The information recorded in the dataset includes Date, Time, Longitude, Latitude, (changes in pitch of the animals' necks (ΔPitch) and side to side neck movements (ΔRoll). Distance, Speed, and Turning Angle were derived from the recorded data. By analysing the dataset, the behaviours of the livestock were classified into grazing, walking, camping-active, and camping-inactive over a period of time.

Finally, in some of the previous work, livestock behaviour classification was not achieved by applying a machine learning technique alone [6, 13, 21]. Those methods also required support of human visual observations. By contrast, the method proposed in this study only relies on the sensor data collected from the livestock, with no requirement for corresponding visual observation by humans. This will make it possible to extend the method to a wider range of grazing situations, and potentially removing the need for validation of behaviours in each scenario, which are complex and variable in terms of both the animal's behaviours and their environment.

2 Background

Machine learning techniques are becoming popular for classifying animal behaviour [6, 8, 15], due to the large amount of data available today. There has been some limitation in the past of using of machine learning technique due to the problem of data collection and obtaining meaningful analysis [6]. William et al. classified cow behaviour into grazing, walking, and resting. The authors have applied four methods to do the behaviour classification, which are JRip, J48, Naïve Bayes, and RF. JRip and RF have shown to provide the best performance for classifying the behaviours. It is also noted that the model predicted the three behaviours using one-level classification method [6].

In a different study, Homburger [10] performed the livestock behaviour classification based on GPS position. They used three algorithms, which are RF, SVM and Linear Discriminant Analysis, to classify three livestock behaviours at one-level, namely grazing, walking, and resting, directly from sensor data. They used the three-axis accelerometers in their research on livestock behaviour. The tracking device is put under livestock jaws and have the potential of classifying more behaviours, such as grazing, lying, running, standing and walking [8]. Given that more behaviours might be able to be deduced from the data collected from the tracking devices, this paper anticipated that a hierarchical classification method could be more accurate.

3 Data and Methodology

3.1 Data

In this paper, the data used for the study was obtained by the Commonwealth Scientific and Industrial Research Organization (CSIRO) in Australia. The data was captured in a field study that was located in the central Wheatbelt of Western Australia, near Tammin (31°30'19.13"S, 117°33'33.82"E), which is about 180 km east of Perth in Western Australia.

Four sheep were selected randomly from the farm and fitted with the WildTrax GPS tracking collars (produced by Bluesky Telemetry Ltd, Aberfeldy, Scotland). The tracking device, attached to the sheep's necks, recorded animal position (Latitude and Longitude) and activities (roll and pitch angle) at 5-min intervals. Due to an error of a tracking device, the data from one sheep has missing values in the time series, so only the data from three sheep can be used for analysis. In line with similar research in the field, the amount of data from the three sheep can be used to validate the establish model [10, 11].

The CSIRO Floreat Laboratory Animal Ethics Committee has approved the protocol for the experimental work undertaken and monitored the welfare of the animals (organisational approval reference #0715).

3.2 Data Collection and Labelling of Data

In many previous studies, livestock behaviours are inferred from the GPS and the sensor data [13, 15, 19, 20]. In this study, two types of data were used. There are the GPS data that has information on the Longitude and Latitude, and the sensor data, which has the Pitch and Roll angles. These two types of data depicted the livestock position and its activities. The Latitude and Longitude data can be used to calculate the distance the livestock move from one point to the next point. Latitude and Longitude values were also used to deduce the livestock turning angle. The Pitch and Roll sensor data indicate the neck movement of the animals.

The distance along the earth's surface can be calculated based on how far apart the two points are. In this study, even though the farmland is relatively flat and the sheep do not walk a long-distance, the haversine formula is applied to compute the distance between two points using the Latitude and Longitude of the two points.

Turning Angle is the measurement of the changing angle of the livestock trajectory from the first position to the third position, where the second position is the position where the livestock may have a change in direction [22]. Figure 1 shows the graphical illustration of how the turning angle is obtained.

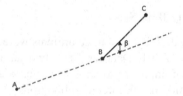

Fig. 1. The animal moves from the first position (A) to the third position (C) via the second position (B). The turning angle is the angle between the dashed line and the thick line. It is symbolised by β. The dashed line connects the first and second positions (A-B). The thick line connects the second position to the third position (B-C).

Another crucial process is to determine the behaviour label. Some studies in livestock behaviour employ an observer to specify livestock behaviour by visual observation. Sometimes the observers may be distracted or fatigued so that their personal condition can affect the observation results. In our study, we undertook a desk evaluation using trajectory analysis. The trajectory of a livestock animal was visualised in a map created from the GPS data from the tracking device. An expert animal observer examined the trajectory map and determined the behaviour at 5-min intervals.

Figure 2 shows the trajectory map with the livestock position in 5-min intervals.

Fig. 2. The map of livestock trajectories. It consists of the livestock location in 5 min intervals for 24 h. The starting point of livestock trajectory was indicated by the red dot. We also created a one-hour livestock trajectory map, and the observer inspected in detail the one-hour duration to determine livestock behaviour. (Color figure online)

3.3   Training and Testing Data Sets

In the data collection stage, data from the three animals were labelled as animal ID280, animal ID285, and animal ID291. Datasets from two animals were combined as a training dataset and dataset from the third animal was used as testing dataset. In order to test the performance of the model, three combinations of the datasets were used to generate the training and testing datasets. Using this way of segregation, we could provide better assessment of the performance of the model. In many instances of prior research in similar application area using machine learning techniques, the available data from all animals are combined into one set [8, 13]. The training and testing sets are normally split into 80% and 20% or 75% and 25% ratio. In this study, it is assumed that the model should be able to predict behaviour in the same flock, and therefore, the data from the three animals are split according to the animal IDs.

Table 1 shows the combination of the datasets from the different animals used to generate the training and testing datasets used in this study.

Table 1. The combination of training and testing dataset

Training data	Testing data
Animal ID280 + Animal ID285	Animal ID291
Animal ID280 + Animal ID291	Animal ID285
Animal ID285 + Animal ID291	Animal ID280

3.4   Second Datasets – Benchmark Dataset

Besides the datasets from CSRIO, we have also used another benchmark dataset obtained from de Weerd et al. [13] to validate the proposed method. In this dataset, it only has the Latitude and Longitude data. The dataset is used to classify seven animal behaviours, which are Grazing, Walking, Drinking, Ruminating, Social, Grooming, and Standing. To be consistent with the dataset obtained from CSRIO, these seven behaviours are then converted to the four main behaviours. Grazing and Walking behaviours were grouped under Active behaviour. Drinking, Grooming and Standing were assigned to Camping Inactive behaviour, and Ruminating and Socialising were assigned to Camping Active behaviour.

3.5   Hierarchical Classification Method

The main purpose of this paper is to propose a hierarchical classification method on livestock behaviours, instead of using a one level classification method used by many prior researchers in the field [8, 9, 13, 14]. The methods proposed in this paper assumed that livestock behaviours can have some form of hierarchical relationship.

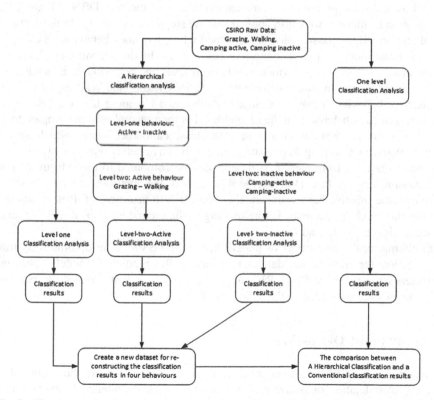

Fig. 3. The conventional one-level vs the proposed hierarchical classification methodology.

Figure 3 shows how the data can be classified into the level one and the level two behaviours.

3.6 Experimental Set Up

In order to demonstrate that the proposed hierarchical classification method can perform well, it is compared to the conventional one-level classification method used by prior research in this field. In the experiments, three classification techniques, which are Deep Belief Network (DBN), Random Forest (RF), and Support Vector Machine (SVM) were used as the classification technique used to identify the livestock behaviours. In order to demonstrate the performance of the proposed hierarchical classification method, it is compared to the conventional one-level classification method used to classify livestock behaviours. Besides comparing the one-level and hierarchical classification methods, the classification performance of DBN, RF and SVM are also evaluated. The following approaches were adopted in the evaluation.

For the conventional one-level classification method, DBN, RF, and SVM classification techniques were applied to each dataset that has four features with label consisting of four behaviours: Grazing, Walking, Camping Active and Camping Inactive. The classification results were calculated by comparing to the ground truth labelled data. For the proposed hierarchical classification method, DBN, RF and SVM classification techniques were also applied and the steps are as follows. In the first step, a level one behaviour classification is performed where livestock behaviour is classified into two behaviours: Active and Inactive behaviours. In the second step, level two behaviour classification is performed, where each level one behaviour is expanded into two more behaviours. In Active behaviour, there are Grazing and Walking behaviours, and Inactive behaviour represent Camping Active and Camping Inactive behaviours. The results of each behaviour in the second level is then recombined to compare to the ground truth labelled data in order to find the classification accuracies, which are also Grazing, Walking, Camping Active and Camping Inactive behaviours. This is to ensure that results obtained from these experiments can be compared to those from the one-level classification method. However, in this method, the level one accuracy based on the Active and Inactive behaviours are also important to take note of. It is desirable to optimise this level one accuracy before moving to the second level, as it will impact on the accuracies in the second level.

Furthermore, the parameters of the DBN, RF, and SVM were optimised by trials and selection for training the dataset to generate the prediction models. The final parameters chosen for the DBN, RF, and SVM techniques that produce the best prediction results are presented in the following Table 2.

4 Results and Discussion

The results for the CSIRO dataset are first presented. Table 3 shows the results of the level one classification of Active and Inactive behaviours generated from the hierarchical method. From the results, it can be observed that the DBN model from each animal ID combination is better when compared to RF and SVM model in training and testing datasets.

Table 2. The model parameters for DBN, RF, and SVM

Parameters	DBN	RF	SVM
Hidden layers	10-16-14-8		
Epoch	300		
Unit function	Maxout – Softmax		
Fine tuning	Backpropagation		
Dropout	0.3		
Learning rate	0.1		
nTree		1000	
mtry		False	
Importance		True	
Gamma			1
Kernel			Sigmoid
Method			SVM Radial

Table 3. The first-level behaviour classification results for hierarchical classification method. The values are in percentage accuracy

Animal ID	Training data			Testing data								
	DBN	RF	SVM	280			285			291		
				DBN	RF	SVM	DBN	RF	SVM	DBN	RF	SVM
280–285	96	95	92	–	–	–	–	–	–	95	95	90
280–291	97	96	94	–	–	–	93	92	86	–	–	–
285–291	95	93	92	98	97	92	–	–	–	–	–	–

The combination of animal ID285 and ID291 model can best classify the behaviours of animal ID280. The model has achieved 98% classification accuracy for Active and Inactive behaviours in the testing data, even though the classification results on the training dataset was the lowest when compared to other animal combinations using DBN techniques.

Imbalance in the class dataset can be observed for level 2 behaviour as only a subset of the data is for level 2. In order to handle the imbalance problem, the SMOTE technique was used to re-sample the training data [23]. Using a new dataset created by SMOTE, the three models using DBN, RF and SVM techniques were used for classifying the second level behaviours. Table 4 shows the classification results for the level two behaviours, grazing and walking. The values are in percentage accuracy.

In the level two Active behaviour, DBN models have given the best prediction accuracy results in the testing set for all animal ID combinations. Similar to the level two Active behaviour, in the level two Inactive behaviour, DBN models also provide better results except the model build from animal ID 280–291 where RF model give the best results, as observed from Table 5.

Table 4. The second-level behaviours, grazing and walking behaviour classification results

Animal ID	Training data			Testing data								
	DBN	RF	SVM	280			285			291		
				DBN	RF	SVM	DBN	RF	SVM	DBN	RF	SVM
280–285				–	–	–	–	–	–	90	85	87
280–291	91	96	88	–	–	–	80	73	76	–	–	–
285–291	88	95	86	84	76	82	–	–	–	–	–	–

Table 5. The level two Camping active and Camping inactive behaviour classification results.

Animal ID	Training data			Testing data								
	DBN	RF	SVM	280			285			291		
				DBN	RF	SVM	DBN	RF	SVM	DBN	RF	SVM
280–285	76	92	81	–	–	–	–	–	–	88	68	76
280–291	85	95	78	–	–	–	66	83	76	–	–	–
285–291	75	96	77	81	78	79	–	–	–	–	–	–

Furthermore, to compare hierarchical classification method and one-level classification method, we report the results using the DBN classification models as this consistently provided better results than RF and SVM. Although we only presented the DBN results here, in all classifiers including those classified using RF and SVM, the hierarchical method is better in accuracy than the one level classification method. Table 6 presents the results of the one level classification method and the hierarchical classification method using DBN technique.

Table 6. The comparison between one-level and hierarchical classification methods using the DBN technique. The values are in percentage accuracy

Animal ID	Training data		Testing data	
	One-level	Hierarchical	One-level	Hierarchical
280285–291	87	94	77	86
280291–285	85	90	83	91
285291–280	83	96	86	93

The comparison results in each animal ID combination showed that hierarchical classification method is better than one level classification method. From the testing results, it can be observed that the hierarchical classification method has better results than one level classification results with the highest accuracy of 93% from animal ID285291 to predict animal ID280.

Like the classification analysis using the CSIRO dataset, the same methodology was applied to the benchmark dataset. In this benchmark dataset, as there is only

Latitude and Longitude features, we therefore only have two variables in the classifier, i.e. distance and turning angle without Pitch and Roll variables. The results are presented in Table 7. The benchmark dataset also show that the use of hierarchical classification method can provide better classification accuracy result when compared to one level classification method.

Table 7. The comparison between one-level and hierarchical classification methods using the DBN technique from the benchmark dataset. The values are in percentage accuracy

Classifiers	Training		Validation	
	One-level	Hierarchical	One-level	Hierarchical
DBN	63	68	66	68
RF	99	99	64	67
SVM	99	99	58	66

From Tables 6 and 7, it can be concluded that the proposed hierarchical classification method consistently provides better behaviour classification accuracies in both the CSIRO and benchmark datasets. From Tables 3, 4 and 5, it can be observed that the DBN technique consistently provided better results than SVM and RF. It could suggest that DBN is a better classification technique to be used to classify livestock behaviours. It is also worth noting from the experiments that the hierarchical approach may reduce data available to the second level and so re-sampling techniques need to be implemented when the training data presented imbalance classes.

5 Conclusion

The hierarchical classification method has been proposed to classify livestock behaviours. The experimental study carried out in this paper is to verify whether the use of the proposed hierarchical classification method can increase classification accuracy when compared to the conventional one level behaviour classification model. The experiments were carried out in two datasets, one collected from CSRIO and another one from benchmark data used in another paper. The results for both datasets showed that hierarchical classification method achieves higher accuracy with all three classification methods, i.e., DBN, RF, and SVM. The study also showed that DBN consistently provide reliable classification accuracies when compared to RF and SVM. This study also demonstrated that the use of SMOTE technique could help to handle the imbalance problem which is common in most data collection.

Acknowledgement. This research was supported by CSIRO Floreat, Western Australia. We are grateful for their cooperation and permission to use their data.

References

1. Manning, L.: What is Ag Big Data? (2015). https://agfundernews.com/what-is-ag-big-data5041.html
2. Carvalho, P.: Can grazing behavior support innovations in grassland management. Tropical Grasslands-Forrajes Tropicales **1**, 137–155 (2013)
3. Rushen, J., Chapinal, N., De Passille, A.: Automated monitoring of behavioural-based animal welfare indicators. Anim. Welfare UFAW J. **21**, 339 (2012)
4. Van Hertem, T., Lague. S., Rooijakkers, L., Vranken, E.: Towards a sustainable meat production with precision livestock farming. In: Proceedings in Food System Dynamics, pp. 357–362 (2016)
5. Manning, J.K., et al.: The effects of global navigation satellite system (GNSS) collars on cattle (Bos taurus) behaviour. Appl. Anim. Behav. Sci. **187**, 54–59 (2017). https://doi.org/10.1016/j.applanim.2016.11.013
6. Williams, M., et al.: A novel behavioral model of the pasture-based dairy cow from GPS data using data mining and machine learning techniques. J. Dairy Sci. **99**, 2063–2075 (2016). https://doi.org/10.3168/jds.2015-10254
7. González, L., Bishop-Hurley, G., Handcock, R., Crossman, C.: Behavioral classification of data from collars containing motion sensors in grazing cattle. Comput. Electron. Agric. **110**, 91–102 (2015). https://doi.org/10.1016/j.compag.2014.10.018
8. Alvarenga, F., et al.: Using a three-axis accelerometer to identify and classify sheep behaviour at pasture. Appl. Anim. Behav. Sci. (2016). https://doi.org/10.1016/j.applanim.2016.05.026
9. Hilario, M.C., Wrage-Mönnig, N., Isselstein, J.: Behavioral patterns of (co-) grazing cattle and sheep on swards differing in plant diversity. Appl. Anim. Behav. Sci. **191**, 17–23 (2017). https://doi.org/10.1016/j.applanim.2017.02.009
10. Homburger, H., Schneider, M., Hilfiker, S., Luscher, A.: Inferring behavioral states of grazing livestock from high-frequency position data alone. PLoS ONE **9**, e114522 (2014)
11. Giovanetti, V., et al.: Automatic classification system for grazing, ruminating and resting behaviour of dairy sheep using a tri-axial accelerometer. Livestock Sci. **196**, 42–48 (2017). https://doi.org/10.1016/j.livsci.2016.12.011
12. Manning, J., et al.: The behavioural responses of beef cattle (Bos taurus) to declining pasture availability and the use of GNSS technology to determine grazing preference. Agriculture **7**, 45 (2017)
13. de Weerd, N., et al.: Deriving animal behaviour from high-frequency GPS: tracking cows in open and forested habitat. PLoS ONE **10**, e0129030 (2015)
14. Diosdado, J.A.V., et al.: Classification of behaviour in housed dairy cows using an accelerometer-based activity monitoring system. Anim. Biotelemetry **3**, 15 (2015). https://doi.org/10.1186/s40317-015-0045-8
15. Wang, G.: Machine learning for inferring animal behavior from location and movement data. Ecol. Inform. **49**, 69–76 (2019). https://doi.org/10.1016/j.ecoinf.2018.12.002
16. Wainberg, M., Alipanahi, B., Frey, B.: Are random forests truly the best classifiers? J. Mach. Learn. Res. **17**, 3837–3841 (2016)
17. Durgesh, K., Lekha, B.: Data classification using support vector machine. J. Theoret. Appl. Inf. Technol. **12**, 1–7 (2010)
18. Hua, Y., Guo, J., Zhao, H.: Deep belief networks and deep learning. In: IEEE International Conference on Intelligent Computing and Internet of Things (ICIT) (2015)
19. Valletta, J., et al.: Applications of machine learning in animal behaviour studies. Anim. Behav. **124**, 203–220 (2017). https://doi.org/10.1016/j.anbehav.2016.12.005

20. Browning, E., et al.: Predicting animal behaviour using deep learning: GPS data alone accurately predict diving in seabirds. Methods Ecol. Evol. **9**, 681–692 (2018). https://doi.org/10.1111/2041-210X.12926
21. Rayas-Amor, A.A., et al.: Triaxial accelerometers for recording grazing and ruminating time in dairy cows: an alternative to visual observations. J. Vet. Behav. Clin. Appl. Res. **20**, 102–108 (2017). https://doi.org/10.1016/j.jveb.2017.04.003
22. Calenge, C., Dray, S., Royer-Carenzi, M.: The concept of animals' trajectories from a data analysis perspective. Ecol. Inform. **4**, 34–41 (2009)
23. Chawla, N., Bowyer, K., Hall, L., Kegelmeyer, W.: SMOTE: synthetic minority over-sampling technique. J. Artif. Intell. Res. **16**, 321–357 (2002)

A Study of Features Affecting on Stroke Prediction Using Machine Learning

Panida Songram[✉] and Chatklaw Jareanpon

Polar Lab, Department of Computer Science, Faculty of Informatics,
Mahasarakham University, Mahasarakham, Thailand
{panida.s,chatklaw.j}@msu.ac.th

Abstract. In 2021, Thailand will become an ageing society. The policy of the health of older people is a challenging task for the Thai government that has to be carefully planned. Stroke is the first leading cause of death of older people in Thailand. Knowing the risk factors for stroke will help people to prevent stroke. In this paper, features affecting stroke are studied based on machine learning. Factors and diseases occurring before stroke are studied as features to detect stroke and find affective factors of stroke. The detection of stroke is investigated based on learning classifiers, SVM, Naïve Bayes, KNN, and decision tree. Moreover, Chi2 is adopted to find affective factors of stroke. The four most affective factors of stroke are focused to know the risk of stroke. From the study, we can see that the factors are more affective than the diseases for detecting stroke and decision tree is the best classifier. Decision tree gives 72.10% of accuracy and 74.29% of F-measure. The factors affecting stroke are smoking, alcohol, cholesterol, blood pressure, sex, exercise, and occupation. Moreover, we found that no smoking can avoid stroke. Drinking alcohol, abnormal cholesterol, and abnormal blood pressure raise the risk of a stroke.

Keywords: Stroke prediction · Stroke classification · Risk factors for stroke · Affective factors of stroke

1 Introduction

Thailand will become an ageing society in 2021 because the number of older people in the country is growing while the proportion of children trends to be decreased. The Thai government plans the development of policies and programming to support older persons. One of the policies is the health of elderly [1]. Older people face special physical and mental health challenges that need to be recognized. Stroke is a top leading cause of death of older people in Thailand and the world. Stroke is a medical emergency that needs immediate medical attention. During a stroke, the brain does not receive enough oxygen or nutrients causing cells to die. Therefore, stroke risk prediction can contribute significantly to its prevention and early treatment. Many researchers tried to study stroke.

© Springer Nature Switzerland AG 2019
R. Chamchong and K. W. Wong (Eds.): MIWAI 2019, LNAI 11909, pp. 216–225, 2019.
https://doi.org/10.1007/978-3-030-33709-4_19

For example, Zhang et al. [2] proposed a risk score derived from Caucasian cohorts to predict stroke. The risk score is calculated using age, blood pressure, smoking, and cholesterol. Chawla et al. [3] proposed a method to detect and classify stroke from brain CT images. The method consists of image enhancement, detection of min-line symmetry and classification of abnormal slices. A two-level classification schema is used to detect abnormalities using features derived in the intensity and the wavelet domain. The method gives high performance in detecting abnormality at the patient levels. Khosla et al. [4] integrate a machine learning approach to predict stroke. The Cardiovascular Health Study dataset is used to study for predicting stroke. First, a systematic method is applied to fill the missing entries in the dataset. Then the relevant features are selected and learned to create classifiers. Two classifiers are constructed by using Support Vector Machines and Margin-based Censored Regression and compared to Cox models. The comparison shows that machine learning methods outperform the Cox model. Gillebert et al. [5] proposed a method to automatically delineate infarct and hemorrhage in stroke CT images. The method performs the accurate normalization of CT images into template space and compares the subsequence voxel-wise with a group of control CT image to identify hypo- or hyper-intense signals. Kansadub et al. [6] adopted data mining techniques to predict stroke disease. Decision Tree, Naive Bayes, and Neural Network are studied to create classifiers for predicting stroke. From the study, Decision tree is the most accurate and Naive Bayes is the best in the area under ROC curve. Mcheick et al. [7] propose a prediction framework based on ontology and Bayesian Belief Networks. The stroke prediction system was proposed to handle the uncertainty of having a stroke disease by determining the risk score level. Singh and Choudhary [8] predict stroke on the Cardiovascular Health Study dataset. First, features are selected by using decision tree and then a classifier is constructed by adopting a back propagation neural network. The patient's physical, mental, blood and diagnosis data are used for stroke prediction.

In this paper, factors and diseases occurring before stroke are studied as features to detect stroke and find affective factors of stroke. We concentrate on both factors and disease because some factors strongly affect strokes and some diseases may lead to stroke. The factors and diseases are investigated for detecting stroke based on machine learning and the factors are ranked to find the most affective factors for stroke.

The rest of this paper is organized as follows: Sect. 2 presents the data collection and preparation used in this work. Section 3 explains the overall methodology for detecting stroke. Section 4 shows experiments and discussion. Finally, the conclusion is provided in Sect. 5.

2   Data Collection and Preparation

Two datasets are collected from a hospital database in Mahasarakham hospital. They are retrieved from records of older people who are over 60 years old. The records are dismissed if some of the factors are missed. Finally, 500 records of stroke and 500 records of non-stroke are collected in this works. Factors and diseases relating to stroke are focused in this paper, so 1,000 records are split into two datasets. The first dataset contains eight factors and classes. Eight factors consist of sex, marry, occupation, exercise, alcohol, smoking, blood pressure, and cholesterol. The second dataset contains diseases occurring before stroke and classes. The records having at least 3 diseases are studied to find diseases that are related to stroke. Due to the variety of diseases, the diseases are grouped to ICD-10 that a medical classification list by the World Health Organization (WHO) [9] as shown in Table 2. The factors are represented in numbers according to Table 1. The diseases are represented by 1 or 0. 1 means that the disease occurs, 0 means that the disease does not occur. y and n represent non-stork and stork classes. The characteristics of the datasets as shown in Table 3. An example of the first and second datasets are shown in Figs. 1 and 2.

Table 1. The meaning of factors

ID	Factor	Meaning
1	Sex	10 = male, 11 = female
2	Marry	20 = marry, 21 = single
3	Smoking	40 = smoke, 41 = used to smoke, 42 = never smoke
4	Alcohol	50 = drink, 51 = used to drink, 52 = never drink
5	Exercise	60 = yes, 61 = no
6	Occupation	700 = village headman, 701 = farmer, 702 = political official
		703 = pensioner, 704 = merchant, 705 = business owner
		706 = police, 707 = priest, 708 = contractor
		709 = employee, 710 = teacher, 711 = local government officer
		712 = hawker, 713 = novelist
7	Blood pressure	80 = abnormal, 81 = normal
8	Cholesterol	90 = abnormal, 91 = normal

Table 2. ICD-10

ID	ICD-10
1	A00-A09 (Intestinal infectious diseases)
2	A15-A19 (Tuberculosis)
3	A20-A28 (Certain zoonotic bacterial diseases)
...	...
227	Z00-Z99 (Factors influencing health status and contact with health services)

Table 3. Characteristics of datasets

Characteristic	Count
The number of factors	8
The number of disease	170
The average length of disease sequence	7
The maximal length of disease sequence	14
The minimal length of disease sequence	3
The average number of patient of a disease	7
The maximum number of patient of a disease	487
The minimum number of patient of a disease	1

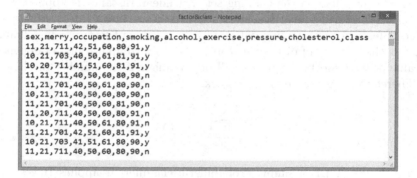

Fig. 1. Factors dataset

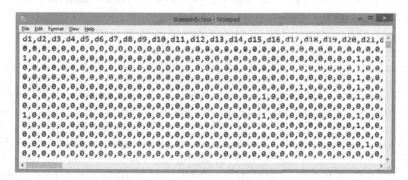

Fig. 2. Diseases dataset

3 Stroke Detection

In this work, factors and diseases occurring before stroke are investigated as features for detecting stroke. First, different classifiers are studied to find the best classifier for detecting stroke. Moreover, features are ranked to find affective factors for stroke.

3.1 Learning Classifiers for Prediction

SVM, Naïve Bayes, KNN, and decision tree are employed to build classifiers for prediction. SVM [10] uses linear or non-linear delineations between classes to divide the data space. It tries to determine the optimal boundaries that maximally split the different classes. In this work, the linear kernel in SVM is used for learning. A new data x can be classified in form as shown in (1).

$$prediction(x) = sgn[b + w^T x] \text{ for } w = \sum_i \alpha_i x_i \qquad (1)$$

KNN [11] is a non-parametric classification method with effective computation for classification. It classifies new data by using the class label of the K most similar neighbor in the training set. Euclidean distance is applied in this work to measure the similarity between two data as shown in (2), where x_i is a value of feature i in a testing data and y_i is a value of feature i in a training data, n is the number of distinct features. To avoid computationally expensive, the value of k should be set to a small odd number if the number of classes is two. Therefore, K is set to 3 in this work.

$$d = \sqrt{\sum_{i=1}^{n}(x_i - y_i)^2} \qquad (2)$$

Naïve Bayes [12] is a simple probabilistic classifier. It applies Bayes' theorem with strong independence assumption between the features. This work uses maximum a posteriori or MAP decision rule. The function assigns a new data $y = \{x_1, x_2, \ldots, x_n\}$ in a class label c_k for k classes as shown in (3).

$$P(c_k \mid y) = \underset{k=\{1,\ldots,k\}}{\arg\max} P(c_k) \prod_{i=1}^{n} P(x_i \mid c_k) \qquad (3)$$

Decision tree [13] is a classifier expressed in a tree of decision. The tree consists of internal nodes and leave nodes. Internal nodes split the instance space into k-sup-spaces accounting to possible attribute values and leave nodes is decision nodes. The decision tree hierarchically classifies the data space into different classes according to the occurrence of terms. The data space is recursively divided until each leaf node contains a class label. Various algorithms proposed for decision tree such as ID3 and C4.5. This work adopts the C4.5 algorithm [14] for learning. It builds the decision tree by using information gain. The feature with the highest information gain is chosen to make the decision.

3.2 Feature Selection

In this paper, features are ranked by using Chi2 to find affective factors. Chi2 [15] is used to test whether the occurrence a specific feature and the occurrence of a specific class are independent. The features are ranked by using a score metric. The score of feature A is calculated by (4), where $C = \{c_1, c_2, \ldots, c_k\}$ is a set of classes, $V = \{v_1, v_2, \ldots, v_n\}$ is a set of possible values of feature A, O_{ij} is the number of observations having the same value v_i in class c_j, R_i is the number of observations having the same value v_i, D_j is the number of observations of class c_j, N is the total number of observations, E_{ij} is the expected frequency of A_{ij}, which can be calculated by $R_i * D_j / N$.

$$\chi^2 = \sum_{i=1}^{n} \sum_{j=1}^{k} \frac{(O_{ij} - E_{ij})}{E_{ij}} \tag{4}$$

3.3   Evaluation

For evaluating the classifiers, 10-fold cross validation is used to divide training and testing sets. Stroke and non-stroke are predicted in a confusion matrix as shown in Table 4, where a represents the number of non-stroke correctly classified, b is the number of non-stroke misclassified as stroke, c refers to the number of stroke misclassified as non-stroke, and d expresses the number of stroke correctly classified. Accuracy, precision, recall, and F-measure are metrics evaluated in our work. Accuracy evaluates the overall predictions of classifier that were correct. It is defined as $Acc = (a + d)/(a + b + c + d)$. Precision is the number of correct predictions from all predictions for each class. Precision for stroke class is defined as $P_{stroke} = d/(b + d)$. Precision for non-stroke class is defined as $P_{non-stroke} = a/(a + c)$. Recall is the number of correct prediction from all true data for each class. Recall for stroke class is defined as $R_{stroke} = d/(c + d)$ and recall for non-stroke class is defined as $R_{non-stroke} = a/(a + b)$. F-measure is the harmonic mean between precision and recall, F-measure for stroke class is defined as $F_{stroke} = 2P_{stroke}R_{stroke}/(P_{stroke} + R_{stroke})$ and F-measure for non-stroke class is defined as $F_{non-stroke} = 2P_{non-stroke}R_{non-stroke}/(P_{non-stroke} + R_{non-stroke})$.

Table 4. Confustion matrix

True	Predicted	
	Non-stroke	Stroke
Non-stroke	a	b
Stroke	c	d

4   Experiments and Discussion

All the experiments are performed using WEKA 3.8 [16]. First, factors are investigated for classifying stroke and non-stroke and resulted in Table 5. From Table 5, we can see that the decision tree is the best classifier for predicting stroke and non-stroke because it performs well with the small number of features and the number of possible values of the feature is limited. It gives the highest accuracy at 72.10%. Moreover, decision tree outperforms SVM, KNN, and Naïve Bayes in terms of F-measures of stroke and non-stroke. The F-measure for stroke is 74.29% and the F-measure for non-stroke is 69.21%. Diseases occurring before stroke are also studied to predict stroke and resulted in Table 6. From Table 6, it shows that the best classifier is SVM due to the high dimension of diseases dataset and SVM performs well in high dimensional space. However, SVM gives accuracy at 62.20% which is less than the accuracies of all classifiers when are evaluated on factors. Therefore, factors are more affective than diseases for predicting stroke. Next, the weights of factors are found and the factors are ranked by using Chi2. The result is shown in Table 7. From Table 7, the merry does not affect stroke and the most affective factor is smoking. Figure 3 shows that non-smokers do not have a risk of stroke while smokers have a higher risk of stroke than non-smokers. Moreover, alcohol is a major contributing cause to stroke as shown in Fig. 4. Furthermore, people who have abnormal cholesterol and blood pressure increase the risk of stroke as shown in Figs. 5 and 6.

Table 5. Classification performance of the learning classifiers on factors

Classifier	Parameter	Accuracy	Non-stroke			Stroke		
			P	R	F	P	R	F
KNN	$K = 3$	70.20	75.52	60.86	66.95	67.06	80.16	72.68
Naive Baye	Maximum a posteriori	70.9	77.34	59.45	66.90	66.97	82.53	73.71
SVM	Linear kernel	69.8	73.73	62.37	67.04	67.59	77.47	71.67
Decision tree	C4.5	**72.10**	77.91	63.44	**69.21**	69.07	81.66	**74.29**

Table 6. Classification performance of the learning classifiers on diseases

Classifier	Parameter	Accuracy	Non-stroke			Stroke		
			P	R	F	P	R	F
KNN	$K = 3$	54.7	53.64	72.63	61.42	57.34	37.16	44.76
Naive Baye	Euclidean distance	58.50	60.83	49.57	54.32	57.07	67.65	61.64
SVM	Linear kernel	**62.20**	60.42	71.10	65.04	67.13	53.72	58.50
Decision tree	C4.5 algorithm	57.50	56.64	65.85	60.30	60.12	50.01	53.79

Table 7. The ranking factors

Rank	Feature description	Weight
1	Smoking	140.25
2	Alcohol	135.39
3	Cholesterol	67.64
4	Blood pressure	65.96
5	Sex	29.76
6	Exercise	18.63
7	Occupation	7.05
8	Merry	0

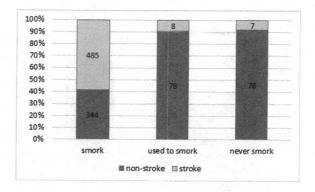

Fig. 3. Distribute of smoking

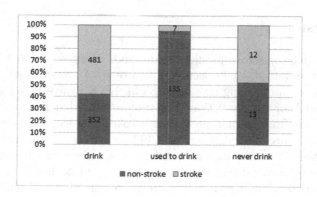

Fig. 4. Distribute of alcohol

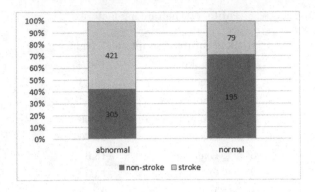

Fig. 5. Distribute of cholesterol

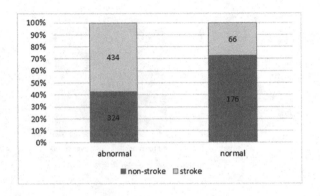

Fig. 6. Distribute of blood pressure

5   Conclusion

The features affecting stroke prediction are studied based on machine learning. The eight factors and diseases occurring before stroke are investigated to predict stroke. Different learning classifiers, SVM, KNN, Naïve Bayes and decision tree, are adopted for the prediction. The results show that the factors are more affective than the diseases for predicting stroke and decision tree is the best classifier. Factors are ranked based on Chi2 to find most affective factors for detecting stroke. From the experimental results, only seven factors, smoking, alcohol, cholesterol, blood pressure, sex, exercise, and occupation, are affective for detecting stroke. The top-4 factors are deeply studied and found that non-smokers do not have a risk of stroke while smokers have the risk of stroke. Moreover, alcohol is a major contributing cause of stroke. Furthermore, people who have abnormal cholesterol and blood pressure increase the risk of stroke.

References

1. Ageing Population in Thailand. http://ageingasia.org/ageing-population-thailand/. Accessed 10 Jan 2019
2. Zhang, X.-F., Attia, J., D'Este, C., Yu, X.-H., Wu, X.-G.: A risk score predicted coronary heart disease and stroke in a chinese cohort. J. Clin. Epidemiol. **58**(9), 951–958 (2005)
3. Chawla, M., Sharma, S., Sivaswamy, J., Kishore, L.T.: A method for automatic detection and classification of stroke from brain CT images. In: Proceeding of 31st Annual International Conference of IEEE Engineering in Medicine and Biology Society (EMBC), USA, pp. 3581–3584. IEEE (2009)
4. Khosla, A., Cao, Y., Chiung-Yu Lin, C., Chiu, H.-K., Hu, J., Lee, H.: An integrated machine learning approach to stroke prediction. In: Proceeding of the 16th ACM SIGKDD International Conference on Knowledge Discovery and Data Mining, USA, pp. 183–192 (2010)
5. Gillebert, C.R., Humphreys, G.W., Mantini, D.: Automated delineation of stroke lesions using brain CT images. NeuroImage Clin. **4**(C), 540–548 (2014)
6. Kansadub, T., Thammaboosadee, S., Kiattisin, S., Jalayondeja C.: Stroke risk prediction model based on demographic data. In: Proceeding of the 8th International Conference on Biomedical Engineering, Thailand, pp. 1–3 (2015)
7. Mcheick, H., Nasser, H., Dbouk, M., Nasser, A.: Stroke prediction context-aware health care system. In: Proceeding of International Conference on Connected Health: Applications, Systems and Engineering Technologies, USA, pp. 30–35 (2016)
8. Singh, M.S., Choudhary, P.: Stroke prediction using artificial intelligence. In: 8th Annual Industrial Automation and Electromechanical Engineering Conference, Thailand, pp. 158–161 (2017)
9. World Health Organization: ICD-10 International Statistical Classification of Diseases and Related Health Problems, 2nd edn. World Health Organization, Geneva, Switzerland (2006)
10. Cortes, C., Vapnik, V.: Support-vector networks. Mach. Learn. **20**(3), 273–297 (1995)
11. Guo, G., Wang, H., Bell, D., Bi, Y., Greer, K.: KNN model-based approach in classification. In: Meersman, R., Tari, Z., Schmidt, D.C. (eds.) OTM 2003. LNCS, vol. 2888, pp. 986–996. Springer, Heidelberg (2003). https://doi.org/10.1007/978-3-540-39964-3_62
12. John, G., Langley, P.: Estimating continuous distributions in Bayesian classifiers. In: Proceeding of the 11th International Conference on Uncertainty in Artificial Intelligence, San Mateo, pp. 338–345 (1995)
13. Quinlan, J.R.: Introduction of decision trees. Mach. Learn. **1**(1), 81–106 (1986)
14. Quinlan, J.R.: C4.5: Programs for Machine Learning, 1st edn. Morgan Kaufmann, San Francisco (1993)
15. Liu, H., Setiono R.: Chi2: feature selection and discretization of numeric attributes. In: Proceedings of the Seventh International Conference on Tools with Artificial Intelligence, pp. 388–391. IEEE Computer Society, USA (1995)
16. Hall, M., Frank, E., Holmes, G., Pfahringer, B., Reutemann, P., Witten, I.H.: The WEKA data mining software: an update. ACM SIGKDD Explor. Newsl. **11**(1), 10–18 (2009)

Short Papers

Short Papers

Content-Based Health Recommender System for ICU Patient

Asif Ahmed Neloy$^{(\boxtimes)}$, Muhammad Shafayat Oshman,
Md. Monzurul Islam, Md. Julhas Hossain, and Zunayeed Bin Zahir

Electrical and Computer Engineering Department, North South University,
Dhaka, Bangladesh
{asif.neloy,shafayat.oshman,monzurul.islam,
julhas.hossain,zunayeed.zahir01}@northsouth.edu

Abstract. In this study, the authors propose a generic architecture, associated terminology and a classificatory model for observing ICU patient's health condition with a Content-Based Recommender (CBR) system consisting of *K-Nearest Neighbors (KNN)* and *Association Rule Mining (ARM)*. The aim of this research is to predict or classify the critically conditioned ICU patients for taking immediate actions to reduce the mortality rate. Predicting the health of the patients with automatic deployment of the models is the key concept of this research. IBM Cloud is used as *Platform as a Service (PaaS)* to store and maintain the hospital data. The proposed model demonstrates an accuracy of *95.6%* from the KNN Basic '*ball_tree*' algorithm. Also, real-time testing of the deployed model showed an accuracy of *87%* while comparing the output with the actual condition of the patient. Combining the IBM Cloud with the Recommender System and early prediction of the health, this proposed research can provide a complete medical decision for the doctors.

Keywords: Content-Based Recommender · KNN · Association mining · Apriori · Eclat · IBM cloud · ICU patient monitoring · Out of sample validation

1 Introduction

Recommender systems fall in the domain of active learning of unsupervised machine learning. The main branches of this learning are – *Collaborative Filtering Methods* and *Content-Based Systems* [1]. The Content-based methods are based on the similarity of item attributes, features, and collaborative methods that calculates similarity from interactions of each attribute. Considering the dataset studied in this research, the authors propose a *Content-Based Recommender System (CBR)*. A survey was conducted before designing the system architecture. The survey contained the information regarding the *Clinical Health Parameters (CHP)* of a patient, current technological aspects, uses in the medical sector and recent advancement in health care. Lack of technology in the field of Patient Care in Bangladesh especially the real-time feedback system for disease diagnostics for ICU Patient is the main problems figured out from the survey. Also, the survey exposed the communication gap between the ICU environment and the on-duty doctor or nurse. Moreover, previous proposed research only

© Springer Nature Switzerland AG 2019
R. Chamchong and K. W. Wong (Eds.): MIWAI 2019, LNAI 11909, pp. 229–237, 2019.
https://doi.org/10.1007/978-3-030-33709-4_20

suggests either hardware mechanisms or machine learning architecture, but no application or implementation regarding this problem has supplied yet. So, a sustainable and scalable ICU monitoring system is still due to date in Bangladesh.

In the current scenario, the acquired dataset for this study requires a large predictive model to run within less complexity. The core purpose of this research is to build an adaptable and scalable ICU patient monitoring system using machine learning. This system will also contain a real-time feedback system. The output of this model classifies whether to take immediate medical attention for a particular patient or not. The IBM Cloud and IBM Watson Studio are used in the proposed methodology for allocating and refactoring the dataset and deploy machine learning models in *passive mode*. Through this *PaaS* platform, the model is delivered for *auto-deployment* [2]. Altogether, the proposed methodology presents a reliable and efficient real-time remote ICU patient health evaluating system that can play a vital role in providing better patient care in developing countries.

2   Related Work

Numerous studies and research had been done previously on recommender systems involving machine learning and data mining along with embedded systems. Those proposed research dealt with different data mining techniques and used those data to build a recommender system. Many of the research proposed embedded systems for real-time feedback.

Sunita et al. [3] proposed a system to use *ARM* algorithm Apriori in course of recommendation system. Based on their previous course history, the proposed system recommends students the type of courses they might be interested in. Santosh et al. [4] classifies the state-of-the-art research in health risk prediction systems or disease prediction systems. The paper showed a detailed literature review of machine learning methods in health risk predictions and also embodied the researches done so far in the fields on health prediction sector. Nithya et al. [5] depicted the study on various prediction techniques and tools for Machine Learning in practice. A glimpse on the applications of Machine Learning in various domains is also discussed in the paper by highlighting its prominent role in the health care industry. Shimada et al. [6] designed a drug recommendation system along with a decision support system that helps doctors to select appropriate first-line drugs. In an evaluation of this prototype system, the risk level it determined by correlating with the decisions of specialists.

Considering the system requirements and current work done in this domain, this proposed study is designed in such a way that, it fulfills the gap of the real-time sustainable feedback system, cost-effective ICU caring system, minimize the communication gap between ICU systems and doctors and most importantly a remote, scalable recommender system achieving the state-of-the-art.

3 Proposed Methodology

The sample dataset is obtained through both hospital registration terminal and central database. The output of the model is stored in the central database through IBM cloud and displayed to the doctor through hospital terminal for examining the model output. ARM plays a key role to endorse the *Frequently Generated Dataset (FGD)* for *Feature Extraction (FE)*. After generating the *FE with FGD* and *Brunching*, KNN is applied to train the whole model. *Out of Sample Validation* and *Evaluation Metrics* precisely evaluate the whole model afterward. Figure 1 is the overview of this *Genetic Architecture* applied in the methodology.

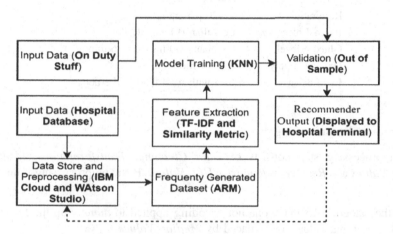

Fig. 1. The architecture of the Proposed System

3.1 Sample Dataset

The *5000-sample* labeled dataset contains *16 features*, among them *10* are *Categorical Features* and *6* are *Numerical Features*. Table 1 illustrates the sample dataset.

Table 1. Sample dataset

Patient No	Age	Primary disease	Internal bleeding	Admitted to	Pulse
01	38	Kidney diseases	No	Word	75
02	13	Accident	Yes	ICU	72
Diastole	B/P	Oxygen Supply	Blood Circulation	Glucose Level	Send SMS
85	High	0	Normal	High	No
80	High	80	Low	Normal	Yes
Ventilator	Sex	Last Feedback	Systole		
No	Male	Normal	125		
Yes	Female	Decreasing	130		

Each of the labels from both datasets defines the condition of the patient, more specifically each of the dataset represents *CHP* of a patient. *CHP* is the main factor that define the health status of a patient. All the standard values of these *CHPs* are presented in Table 2 and considered for all age and sex. A common way to measure the condition is to compare the functional status of the parameters with the model output.

Table 2. Standard health parameters of human body

Health parameters	Standard/Normal measurement
Blood Pressure (BP)	120–80 mm Hg
Pulse	60–100 bpm
Respiration	12–20 per minutes
Internal bleeding	Less than 0.1 mg/dL
Glucose level	less than 100 mg/dL (5.6 mmol/L)
Oxygen supply	20% or more
Fluid balance	1:15 with total fluid in the body

3.2 Dataset Preprocessing

Data preprocessing step consists of *Data Cleaning, Transformations, Identifying Missing Values* and *Replace* or *Remove NULL Values*. Basic preprocessing is described below -

- For the categorical values, one hot encoding applied to *dummy* up the features.
- Null or missing values are replaced by *Median Values*.
- Create grouping features for better visualization. The features shown in Table 3 is selected by *Grouping* similar features. The total number of data points and features after grouping is *(3000, 5)*.

Table 3. New Feature Extraction

Previous features	New features
Age, sex, weight, admitted to	User details
Ventilator, internal bleeding, oxygen supply, last feedback	Medical parameters
Systole, diastole, glucose level	Body parameters

- *MinMaxScaler* derived from Eq. (1) is applied to diverse datapoints values to 0 and 1 [7]. In Eq. (1), x is the input data that will be transformed, x_{min} is the *min value* in that particular label, x_{max} is the *maxi value* in the label.

$$X_{sc} = \frac{x - x_{min}}{x_{max} - x_{min}} \tag{1}$$

3.3 Apriori and Eclat

Apriori works by identifying a particular characteristic of a data set and attempting to note how frequently that characteristic pop up throughout the set. Apriori is defined by *Support, Confidence, Lift* and *Conviction* [8]. Figure 2 shows the general process of the Apriori algorithm for dataset examined in this paper.

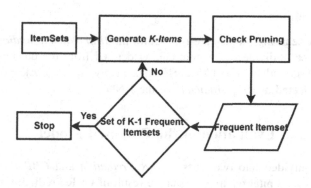

Fig. 2. Apriori data process flowchart for the selecting key phases, provided by similarity metric and TF-IDF for selecting support, confidence, and lift of Apriori algorithms.

3.4 TF-IDF and Similarity Metric

Term Frequency (TF) and *Inverse Document Frequency (IDF)* are used in *Information Retrieval System* that is proposed in this methodology. The *DOT Product* of TF-IDF generates a *Cosine Similarity Metric (CSM)*. This metric says how related are two features by looking at the angle instead of magnitude [9]. In this study, *CSM* selects the *CHP* from the label of the dataset which is more important for a patient. Therefore, this metric is very important for the model to select the adequate features. A sample metric which has been proposed is presented in Table 4. The values of Table 4 are generated from Eq. (2) and using *FGD along with FE* presented in Table 3.

Table 4. Result of *CSM*

	Patient No	Send SMS	User details	Medical parameters	Body parameters
Patient No	1	0.75	0.62	0.57	0.55
Send SMS	0.35	1	0.37	0.33	0.27
User details	0.15	0.16	1	0.13	0.10
Medical parameters	0.05	0.08	0.05	1	0.06
Body parameters	0.02	0.01	0.05	0.04	1

$$w_{t,d} = \begin{cases} 1 + \log_{10} tf_{t,d}, & \textit{if } tf_{t,d} > 0 \\ 0, & \textit{otherwise} \end{cases} \text{ and } w_{t,d} = tf_{t,d} * \log \frac{N}{D_{f,t}} \qquad (2)$$

here, $tf_{t,d}$ is the number of occurrences of t in dataset d. $D_{f,t}$ is the number of features containing the term t. N is the total number of features in the label (health parameters of the patients).

3.5   Model Training

All the machine learning models are run within the *IBM Watson studio*. The Watson Studio automatically divides the whole dataset derived from the database into *Train, Test, Validation* set with a ratio of **70:20:10** respectively. $n_{neighbors}(k) = 6, algorithm = $ '*ball_tree*' is selected as the *parameters* of the KNN.

4   Analysis and Evaluation of the Proposed Model

The results are divided into two parts – *CBR Prediction* and *CBR Prediction Comparison* from live hospital reading. A sample result of CBR Prediction obtained from the KNN model is presented in Table 5.

Table 5. CBR results from KNN

Patient No	Normal score	Normalized rank	Normalized score	IDF score
2877	**7.3**	**9**	**1.696**	**0.36**
168	6.8	14	1.368	0.32
2	6.0	23	1.023	0.29
1533	5.6	87	0.896	0.26
66	5.1	103	0.562	0.23

In Table 5, *Normal Score* indicates the *Patient ID* which is obtained from the KNN based on the *ball_tree* algorithm. The *IDF Score* is obtained from *TF-IDF* to secure the *Wrong Predicted Vectors* [10]. Comparing with the CHP (Table 2), *Normal Score* more than **5.00**, *Normalized Score* more than **0.50** and *IDF Scores* more than **0.20** are considered as the *Critical Score* for a patient. Furthermore, the missing important rating score was calculated through the *Normalized Score*. It is observed by comparing the *FE*. This score refers to the values of the selected features from the *CSM* (Table 4). All the rank is associated with the obtained output from the KNN algorithm. *Higher Normalized Rank* is prior subject to take immediate medical attention.

4.1   Out of Sample Validation

An *out-of-sample validation* is proposed in this paper for validating the model. This method can be described as-withhold some of the sample data from the model while identification and estimation process [11]. After that, use the model to make predictions for the withhold data in order to see how accurate they are. Also, this process determines whether the statistics of their errors are similar to those that the model made within the sample of data that was fitted [11]. This overall scenario is presented by a *Cross-Validation Curve* with $k = 1\ to\ 30$ and Error estimation using *Mean Square Error (MSE)*, proposed by Kanjanatarakul et al. [12] (Fig. 3). k value 6–13 shows the highest accuracy. MSE score also minimizes the error from k = 6 to 13. Highest accuracy achieved for k = 6 from Cross-validation is **97%** and the lowest error rate at that node is **0.01**.

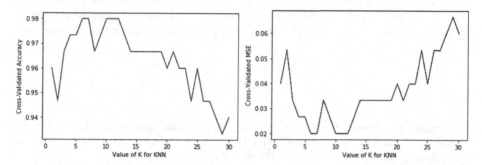

Fig. 3. Validation curves in terms of k values (Left) and MSE error rate (right)

4.2 Validation of the CBR from Hospital Terminal

Other than the Out of Sample Validation process, the model's output is compared through real-time observation from the hospital. In 12 h of observation, a total number of *898 Health Predictions* from *120 patients* were correct among *1025 predictions* (**87% accuracy**). The live test is examined with the *Manual Prediction* of standard *CHP* with the prediction acquired from the deployed KNN model. A sample result from this observation is presented in Table 6. The whole process is tested through the Watson Studio *Performance Monitoring Gateway* [13]. Through this method, the results are updated in every *60* s.

Table 6. CBR result comparison with live observation

Normalized rank	CBR output condition	Normal score	Actual condition observed
98	Increasing	3.2	Normal
86	Normal	4.1	Normal
5	**Critical**	**7.5**	**Decreasing**
2	**Decreasing**	**7.9**	**Decreasing**

5 Conclusion

A content-based recommender system using *KNN, Apriori* and *Eclat* along with *IBM Cloud* as *PaaS* is proposed by the authors. Some of the existed techniques and technologies are used to give a new shape in the hospitals, which has the potential for a large impact on the health care system of developing countries. The auto deployable machine learning models within IBM Watson Studio adds a new dimension in machine learning model building. Another key concept of this research is the parameter tuning and use different validation approaches for the KNN model. The model presented in this paper exhibited an accuracy of more than **92%** with **87%** practically compared result. This research can be enhanced by adding more data points and full health parameters to measure the human body circulations. In the future, an embedded system will be introduced along with the machine learning model to take a live reading from ICU machines.

References

1. Pazzani, M.J., Billsus, D.: Content-based recommendation systems. In: Brusilovsky, P., Kobsa, A., Nejdl, W. (eds.) The Adaptive Web. LNCS, vol. 4321, pp. 325–341. Springer, Heidelberg (2007). https://doi.org/10.1007/978-3-540-72079-9_10
2. Breiter, G., et al.: Software defined environments based on TOSCA in IBM cloud implementations. IBM J. Res. Deve. **58** (2014). https://doi.org/10.1147/JRD.2014.2304772
3. Aher, S.B., Lobo, L.: Combination of machine learning algorithms for recommendation of courses in E-Learning System based on historical data. Knowl. Based Syst. **51**, 1–14 (2013). https://doi.org/10.1016/j.knosys.2013.04.015
4. Shinde, S.A., Rajeswari, P.R.: Intelligent health risk prediction systems using machine learning: a review. Int. J. Eng. Technol. **7**, 1019 (2018). https://doi.org/10.14419/ijet.v7i3.12654
5. Nithya, B., Ilango, V.: Predictive analytics in health care using machine learning tools and techniques. In: 2017 International Conference on Intelligent Computing and Control Systems (ICICCS) (2017). https://doi.org/10.1109/ICCONS.2017.8250771
6. Shimada K., et al.: Drug-recommendation system for patients with infectious diseases. In: AMIA Annual Symposium, p. 1112 (2005)
7. Varoquaux, G., Buitinck, L., Louppe, G., Grisel, O., Pedregosa, F., Mueller, A.: Scikit-learn. GetMobile Mob. Comput. Commun. **19**, 29–33 (2015). https://doi.org/10.1145/2786984.2786995
8. Borgelt, C., Kruse, R.: Induction of Association Rules: Apriori Implementation. Compstat, 395–400 (2002). https://doi.org/10.1007/978-3-642-57489-4_59
9. Lahitani, A.R., Permanasari, A.E., Setiawan, N.A.: Cosine similarity to determine similarity measure: study case in online essay assessment. In: 2016 4th International Conference on Cyber and IT Service Management (2016). https://doi.org/10.1109/citsm.2016.7577578
10. Qaiser, S., Ali, R.: Text mining: use of TF-IDF to examine the relevance of words to documents. Int. J. Comput. Appl. **181**, 25–29 (2018). https://doi.org/10.5120/ijca2018917395
11. Pang, H., Jung, S.-H.: Sample size considerations of prediction-validation methods in high-dimensional data for survival outcomes. Genet. Epidemiol. **37**, 276–282 (2013). https://doi.org/10.1002/gepi.21721

12. Kanjanatarakul, O., Kuson, S., Denoeux, T.: An evidential K-nearest neighbor classifier based on contextual discounting and likelihood maximization. In: Destercke, S., Denoeux, T., Cuzzolin, F., Martin, A. (eds.) BELIEF 2018. LNCS (LNAI), vol. 11069, pp. 155–162. Springer, Cham (2018). https://doi.org/10.1007/978-3-319-99383-6_20
13. Cecil, R.R., Soares, J.: IBM Watson studio: a platform to transform data to intelligence. In: Barbosa-Povoa, A.P., Jenzer, H., de Miranda, J.L. (eds.) Pharmaceutical Supply Chains - Medicines Shortages. LNL, pp. 183–192. Springer, Cham (2019). https://doi.org/10.1007/978-3-030-15398-4_13

Domain-General Versus Domain-Specific Named Entity Recognition: A Case Study Using TEXT

Cheng Yang Lim[1], Ian K. T. Tan[2](✉) (iD), and Bhawani Selvaretnam[3]

[1] Faculty of Computing and Informatics, Multimedia University,
63100 Cyberjaya, Selangor, Malaysia
cylim.pg@gmail.com
[2] School of IT, Monash University Malaysia, Bandar Sunway,
47500 Subang Jaya, Selangor, Malaysia
ian.tan1@monash.edu
[3] Valiantlytix Sdn Bhd, Pinnacle Petaling Jaya, Jalan 51a/223, PJS 52,
46100 Petaling Jaya, Selangor, Malaysia
bhawani@valiantlytix.com

Abstract. Named entity recognition (NER) seeks to identify and classify named entities within bodies of text into language categories such as nouns, that are reflective of locations, organizations, and people. As it is language dependent, the approach taken for most NER systems are domain-general, meaning that they are designed based on a language and not on a specific targeted domain. With current usage of non-formal languages on social media, this instigates the need to compare the performance of domain-general and domain specific NERs. A domain specific NER (vehicle traffic domain), TEXT, is described and the performance of domain-general NER versus TEXT is compared. The results of the evaluation show that the performance of domain-specific NER significantly outperforms domain-general NER. The domain-general NER could only perform adequately for common scenarios.

Keywords: Domain-general · Domain-specific · Named Entity Recognition · Traffic · Information extraction

1 Introduction

The recent renewed interest in machine learning has been spurred by a combination of high performance computing and affordable fast storage, that are fueled by matured as well as new machine learning algorithms.

In the last few years, the term machine learning has been marketed to us that it utilizes advanced techniques that are more powerful than what we have been exposed to in the past. However, the question arises when majority of the large corporations [4] are still implementing the traditional rule-based approaches.

© Springer Nature Switzerland AG 2019
R. Chamchong and K. W. Wong (Eds.): MIWAI 2019, LNAI 11909, pp. 238–246, 2019.
https://doi.org/10.1007/978-3-030-33709-4_21

If the promises of machine learning are as expected, this debate on how ready is machine learning needs to be explored.

Of interest in the recent years has been the application of machine learning for text processing in the area of social media. The sheer amount of data that social media processed has been ascertained through the many use cases of obtaining opinions and sentiments from them.

This paper is to discuss the differences of the machine learning approaches against that of a rule-based approach for Named Entity Recognition (NER) in situations where the aspect is unknown. In noting that GDELT project is a domain-specific project in text processing, we intend to further study using a domain-specific NER versus domain-general NERs using traffic information as the use case. We selected 5 domain-general NERs that are popular among the NER systems and evaluate their performance against TEXT [8], an NER that was designed to extract location and traffic state. Traffic domain is chosen as the area of research as it is a common event reporting. The NER systems that will be used to compare in this paper are: Polyglot [1], OSU [10], MITIE [6], StanfordNER [9], and NLTK [2].

The structure of the paper is as follows: Sect. 2 talks about TEXT, Sect. 3 literates all the NER systems that are used for the evaluation; Sect. 4 explains the experimental setup and the results; Sect. 5 concludes the research done.

2   TEXT

TEXT [8] is a domain-specific NER system developed to extract location and traffic state entities from traffic tweets. Despite the growing popularity of machine learning techniques in NER, TEXT has taken the rule-based approach. Chiticariu et al. [4] stated the significance of the rule-based approach in the industry, proving that the value of rule-based approaches have not diminished. This has validated the decision for taking the rule-based approach in the implementation of TEXT. In addition, TEXT also took the rule-based approach because of the need for a massive training data set for a machine learning (ML) approach and the ability for a rule-based approach to trace any errors. The authors also claimed that it is not economical to annotate a large set of data just for a domain-specific NER. Unlike domain-general NER systems, a domain-specific NER is generally only addressing a smaller problem data set. As such, TEXT has taken rule-based approach.

TEXT operation consists of 4 main algorithms, designed with injection approach. The 4 algorithms were crafted based on the analysis of 3,000 random tweets among 65,000 tweets that were obtained. These algorithms utilize special words types, Part-of-Speech (POS) tags, and text positions to capture the entities. The algorithms were designed such that the special word types could be changed, and it could be reused in other cases. The special word types are described in Table 1.

Table 1. Special word types and its descriptions

Type of words	Description	Sample words
Trigger words	Words that triggers the rules	to, from, near
Counter trigger Words	Words that nullify the triggering of the rules	due, available
Boundary words	Words that marks the ending of the entity	to, till, in
Blocking words	Extension or description of the words that usually stand in between of trigger words and the desired entity	the, almost
Extension words	Words that further describes the location	highway, exit

---

Algorithm 1. Using trigger words to detect location

$T \leftarrow$ tokenized tweet text, $POS_T \leftarrow$ POS Tag of tokenized tweets, $tWords \leftarrow$ lists of trigger words, $cWords \leftarrow$ lists of counter trigger words, $bWords \leftarrow$ lists of boundary words, $bkWords \leftarrow$ Lists of blocking words, $eWords \leftarrow$ extension words
$pointer = 0$
for all elements $\in tWords$ **do**
    if $tWords \in T$ **then**
        if $cWords \in n \pm 2$ **then**
            break;
        end if
    end if
    if $n + 1 \in bkWords$ **then**
        $pointer{+}{+}$;
    end if
    if $POS\_T[n + pointer] = NNP$ **then**
        while $POS\_T[n + pointer] = NN$ **do**
            Append $T[n + pointer] \rightarrow tempLoc$;
            $pointer{+}{+}$;
        end while
    else if $bWords$ exists & distance between $tWords < 5$ **then**
        while $n + pointer$ in range & $T[n + pointer] \neq bWords$ **do**
            Append $T[n + pointer] \rightarrow tempLoc$;
            $pointer{+}{+}$;
        end while
    end if
    if len(tempLoc)< 5 **then**
        if $T[n + pointer + 1] \in eWords$ **then**
            $tempLoc \mathrel{+}= T[n + pointer + 1]$;
        end if
        $L \longleftarrow tempLoc$;
    else
        postProcess($tempLoc$);
    end if
end for

Algorithm 1, which is using trigger words and POS tags to detect the location entity within a traffic related text. The algorithm first looks for trigger words in each segment, if any trigger words are present, the algorithm will check if there are any counter-trigger words in the vicinity of the trigger word. If there are no counter trigger words, then the extraction process will be initiated.

Algorithm 2. Using POS and position to detect location

$W \leftarrow$ lists of partial words, $T \leftarrow$ tokenized text, $P \leftarrow$ POs tag of tokenized text
At position 0:
if $T[n] = NNP$ or $T[n] \in W$ **then**
    $n{+}{+}$;
    while P[n] = NN **do**
        tempLoc $\leftarrow$ T[n];
    end while
    L $\leftarrow$ tempLoc;
end if

Algorithm 2, utilizes POS tags and position to detect the location entities. The author claims that most of the location entities (subject) appears at the start of each segment, thus an algorithm was crafted to extract the entities at position 0.

Algorithm 3. Using trigger words to detect state

$tWords \leftarrow$ dictionary of trigger words with respective length (W, m), $T \leftarrow$ tokenized text, $bkWords \leftarrow$ Lists of blockage words
if $T[n] \in tWords$ **then**
    if $T[n + 1] \in bkWords$ **then**
        $n{+}{+}$;
    end if
    S $\leftarrow T[n + 1 : n + m]$;
end if

Similar to Algorithm 1, Algorithm 3 utilizes trigger words to extract state entities.

Algorithm 4. Using POS tag patterns to detect state

---

statePOS $\leftarrow$ lists of POS combination that indicates the state, $T \leftarrow$ tokenized text, $P \leftarrow$ POs tag of tokenized text
if statePOS $\subset$ P **then**
    S$\leftarrow$ T[n+1:n+m];
end if

Algorithm 4 uses a pre-defined POS patterns to extract state entities. The POS patterns were crafted by analyzing the surrounding words around the state entities. Further details of the algorithms are explained in [8].

3   Domain General NER

Polyglot [1] is a multilingual Named Entity Recognition system which supports entity extraction for 40 languages. It was built using Wikipedia and Freebase. Their research addresses the drawbacks of other system by relying on language-independent techniques. In their paper, they addressed and tackles 2 major issues which are: (i) multilingual NER requires large human annotated datasets which are rare; (ii) to be able to perform multilingual NER, designing relevant features and sufficient linguistic proficiency is required for each language of interest. In light, Neural word embeddings, Wikipedia link structure and Freebase attributes were used to automate the construction of the NER annotators for all 40 languages. The system learns neural word embeddings which encode semantic and syntactic features of words in each distinct language and utilizes internal links in Wikipedia articles to detect named entity mentions. Anchor text is included as a positive training example, when a certain link leads to article that was identified by Freebase as an entity article. They implemented Polyglot embeddings as their sole feature for each language and modeled NER as a word level classification problem. They obtained their training corpus by utilizing Wikipedia and Freebase, and cope with missing annotation problem through oversampling and extending annotation coverage. For evaluation, they have evaluated the system's performance on human annotated datasets and for languages where no gold-standard benchmarks are available, they used distant evaluation that implements statistical machine translation. The system can be downloaded by using python pip.

MITIE [6] is a free-ware and is developed by Massachusetts Institute of Technology (MIT). It was built on top of dlib (a machine learning library). It implements distributional word embeddings and Support Vector Machine (SVM) to perform various Natural Language Processing (NLP) task. The system supports across multiple languages, which includes English, Spanish and German.

Standford CoreNLP [9] is a toolkit which is written in Java and provides a set of human language technology tools. Stanford CoreNLP provides a series of state-of-the-art NLP tools which includes but not limited to: POS tagging, NER, dependencies parsing, coreference, POS tagging. Among them, Stanford NER is one of the tools provided. Stanford NER, also as known as CRFClassifier [5] which implements the linear chain Conditional Random Field (CRF) sequence model as the base of the entity recognition process. They claimed that their system is able to achieve a reduction up to 9%, the state-of-the-art system at that current time.

OSU [10] is a NER that utilizes a modified NLP pipeline. The pipeline consists of the following processes: OS tagging, chunking, capitalization, segmentation and entity classification. In their research, they also evaluate the performance of off-the-shelf news trained NLP tools when applied on Twitter tweets.

To increase the performance of the NER, they have redesigned and trained a Part of Speech tagger (TwitterPOS). OSU performs entity extraction on 2,400 tweets with 10 types of entities which are both popular in Twitter and have a good coverage in Freebase [3]. OSU's performance is evaluated against the dataset. The entities that were covered by OSU includes PERSON, GEO-LOC, COMPANY, FACILITY, PRODUCT, BAND, SPORTSTEAM, MOVIE, TV-SHOW, OTHER. In their paper, it is claimed that OSU's classification system outperforms the baselines of Freebase and achieves a 25% increase in F1 score.

NLTK [2] is a Natural Language Toolkit that provides ready-to-use computational linguistics course ware. NLTK comes with various NLP tools which include but not limited to: POS tagging, chunking, chinking, stemming, NER, parsing. For their NER, they first perform segmentation on the text, then tokenize, POS tagging then perform entity detection. The basic technique that were used for entity detection is chunking, which combines multiple tokens and label the sequences together. The chunking utilizes the BIO tags, which are the Begin, Inside and Outside tags to chunk the tokens. The chunks were then used to determine if they are a Named Entity. The performance of the system is not evaluated in their paper.

All the NER system mentioned above are designed for general purpose entity extraction. A summary of the systems are tabulated in Table 2. None of them are domain specific system except for TEXT, which is an NER system that was created exclusively for detecting location and state entities from traffic-related tweets. TEXT utilizes rule-based approach and formulates 4 general rules that utilizes special words to trigger, halt and extend the entity recognition process.

Table 2. Summary of features of each NER system

System	Approach	Type	Domain	Language	Code
Polygot	Neural network	Normal	General	Multiple	Python
NLTK	CRF	Normal	General	English	Python
StanfordNER	CRF	Normal	General	Multiple	Java
MITIE	SVM	Normal	General	English	C++
OSU	SVM	Twitter	General	English	Python
TEXT	Rule Based	Twitter	Traffic	English	Python

4 Experimental Setup

A total of 65,000 unfiltered tweets were collected using Tweepy. From the tweet, 1,500 English tweets and 1,500 Malay tweets are selected. The tweets were selected randomly from the 65,000 tweets. The Malay tweets were translated to English by using a statistical translation machine, MOSES [7], as TEXT only supports the English language. The dataset ranges over 10 days and consists both English and Malay language tweets. Despite the tweets being written in English, most of the location names are in the Malay language.

For the ground truth, each tweet is manually annotated 3 times by distinct annotators and then compiled. If there are any differences in the 3 annotated data, the majority answer will be accepted, and if all 3 answers are different, a mediator will decide which answer to accept. The dataset contains around 7002 location entities and 3752 state entities. The entities are not unique and there are duplicates.

5   Results

To compare the performance of all NER, all the systems were downloaded and setup respectively. They were used to perform entity extraction on the tweets. Due to the limitation of categories that can be extracted, evaluation was performed for extracting the location entity.

For comparison purposes, the traffic dataset was used on all the systems. The results are as tabulated in Table 3.

Table 3. Results of comparison

System	Precision	Recall	F1 Score
Polyglot	55.48	25.07	34.54
NLTK	67.57	11.87	20.19
StandfordNER	83.24	29.47	43.53
MITIE	86.37	38.90	53.64
OSU	87.18	15.91	26.91
TEXT	**90.78**	**91.79**	**91.28**

From Table 3, TEXT outperforms the other domain-generic NER, showing the value of rule-based approach for NER to achieve better results when it is applied on a domain-specific area. The results have proven that domain-specific NER has a better advantage over domain-general NER as domain-specific system is designed solely to solve that certain problem. Unlike domain-general system, the scope of domain-specific system is narrow, allowing it to be less prone to error, easier to design and develop, and easier to trace the source of error. In addition, domain specific system could implement rule-based approach easily due to the ease in formulating rules.

The lower performance of the other NER also can be explained by the training data used to train the system. The domain-general systems are all trained on general data, causing the system to unable to cover all the domains, leading to insufficient training data when applied on just one specific domain. This might be the cause on why TEXT could outperform the domain-general NERs when applied on traffic tweets. On the contrary, domain-specific system might not perform well when it is applied to general data due for the same reason.

In addition, in terms of comparing domain general and domain specific NER system, TEXT also clearly outperforms OSU (a domain general Twitter based NER). However, if we look at the results closely, we could realize that even though the recall is low, the precision is acceptable. The reason this phenomenon occurs might be the effect of the NER systems are trained in general areas, causing the system to only touch the surface of each domain. They could perform well if the text was written in a "simple" way, causing them to be correct from the extractions (higher true positive). In contrast, when the text was written in a more complex or the writing style is not by people who are familiar with the domain, it fails to perform, thus causing a drop in the recall.

6 Conclusion

In this paper, the performance of domain-general NER and domain-specific NER are evaluated and compared. From the results, it is clear that the domain specific NER, TEXT greatly outperforms the other domain-general NER when applied to data on its' domain. From there, we could conclude that domain specific systems can perform better when it is applied on its domain, however, it could perform poorly when it is applied on data out of its domain. The pros and cons of adopting either approach are clear, the decision of adopting which approach lies on what is the scope of the system should perform.

References

1. Al-Rfou, R., Kulkarni, V., Perozzi, B., Skiena, S.: POLYGLOT-NER: massive multilingual named entity recognition. In: Proceedings of the 2015 SIAM International Conference on Data Mining, pp. 586–594. SIAM (2015)
2. Bird, S., Loper, E.: NLTK: the natural language toolkit. association for computational linguistics. In: Proceedings of the ACL Demonstration Session, pp. 214–217 (2004)
3. Bollacker, K., Evans, C., Paritosh, P., Sturge, T., Taylor, J.: Freebase: a collaboratively created graph database for structuring human knowledge. In: Proceedings of the 2008 ACM SIGMOD International Conference on Management of Data, pp. 1247–1250. AcM (2008)
4. Chiticariu, L., Li, Y., Reiss, F.R.: Rule-based information extraction is dead! long live rule-based information extraction systems! In: Proceedings of the 2013 Conference on Empirical Methods in Natural Language Processing, pp. 827–832 (2013)
5. Finkel, J.R., Grenager, T., Manning, C.: Incorporating non-local information into information extraction systems by Gibbs sampling. In: Proceedings of the 43rd Annual Meeting on Association for Computational Linguistics, pp. 363–370. Association for Computational Linguistics (2005)
6. King, D.E.: Dlib-ml: a machine learning toolkit. J. Mach. Learn. Res. **10**, 1755–1758 (2009)

7. Koehn, P., et al.: Moses: open source toolkit for statistical machine translation. In: Proceedings of the 45th Annual Meeting of the Association for Computational Linguistics Companion Volume Proceedings of the Demo and Poster Sessions, pp. 177–180 (2007)
8. Lim, C.Y., Tan, I.K., Selvaretnam, B., Howg, E.K., Kar, L.H.: Text: Traffic entity extraction from Twitter. In: Proceedings of the 2019 5th International Conference on Computing and Data Engineering, pp. 53–59. ACM (2019)
9. Manning, C., Surdeanu, M., Bauer, J., Finkel, J., Bethard, S., McClosky, D.: The stanford coreNLP natural language processing toolkit. In: Proceedings of 52nd Annual Meeting of the Association for Computational Linguistics: System Demonstrations, pp. 55–60 (2014)
10. Ritter, A., Clark, S., Etzioni, O., et al.: Named entity recognition in tweets: an experimental study. In: Proceedings of the Conference on Empirical Methods in Natural Language Processing, pp. 1524–1534. Association for Computational Linguistics (2011)

Pixel-Level Crack Detection in Images Using SegNet

Chunge Song[1], Lijun Wu[1]([✉])(iD), Zhicong Chen[1](iD), Haifang Zhou[1], Peijie Lin[1], Shuying Cheng[1], and Zhenhui Wu[2]

[1] College of Physics and Information Engineering, Fuzhou University,
Fuzhou 350116, China
lijun.wu@fzu.edu.cn

[2] State Grid Fuzhou Electric Power Supply Company, Fuzhou 350116, China

Abstract. Crack detection is a critical task in routine inspection of building structures. Most of the traditional crack detection methodologies are conducted by human inspectors that may submit inaccurate damage assessments. In recent years, deep learning has produced extremely promising results for various tasks. In this work, a lightweight end-to-end pixel-wise classification architecture called SegNet is employed to segment the structure surface cracks. Compared with other semantic segmentation architectures, SegNet uses pooling indices calculated in the pooling step of the encoder to perform nonlinear upsampling in the corresponding decoder, which doesn't require to learn in the upsample. In this paper, a crack image dataset collected under a variety of complex environment are utilized to train and test the SegNet model, i.e. a self-labeled dataset with 2068 bridge cracks images at the size of 1024×1024. In order to improve the generalization ability of network data augmentation is used. The experimental results show that the SegNet outperforms the traditional edge detection algorithm, such as Canny and Sobel, in the dataset. The trained SegNet model can be used to segment the cracks in images at any size with the assistant of sliding window scanning technique.

Keywords: Crack detection · Deep learning · SegNet · Semantic Segmentation

1 Introduction

The presence of surface cracks offers an indication of the stress endured inside the structures. Presently, crack assessment is mainly conducted through labor-intensive manual inspection, which is prone to subjected to the inspector's experience and skill. Several methodologies have been proposed to address the low efficiency and high cost problems in the manual inspection. For example, a laser testing crack detection technique has been developed for use in corrosion fatigue testing of fine metallic wires [1]. It can detect tiny cracks on small specimens, but

R. Chamchong and K. W. Wong (Eds.): MIWAI 2019, LNAI 11909, pp. 247–254, 2019.
https://doi.org/10.1007/978-3-030-33709-4_22

laser detection equipment is expensive. Tong et al. use acoustic emission technology to detect cracks on welded construction of large-scale pressure vessels [2], and then locates the crack accurately through an ultrasonic testing technology. This method can track and monitor to the development of cracks online, but the crack detection result is not good on small-sized equipment. Several crack detection methodologies based on digital image processing technology have also been proposed. For example, based on grayscale threshold segmentation methods [3], edge detection methods [4] and texture feature [5], cracks are successfully separated from the backgrounds. However, the performance of these methodologies is susceptible to the noise in the complex background and the changes of illumination.

In recent years, convolutional neural network (CNN) has achieved great success in various image processing tasks, object detection and identification [6], whole image classification [7] and semantic segmentation [8]. Based on convolutional neural network, Wang et al. [9] successfully classified the sub images into two cartographies, i.e. with or without cracks. However, such image classification algorithm can only locate cracks at patch-level but not at the pixel-level. To address this issue, Zhang et al. [10] proposed a high accuracy pixel-wise crack detection architecture, which is called CrackNet. It is worth to noting that CrackNet does not have any pooling layers, which ensure the size of output image remains the same as the one of input image. However, a predesigned line filters is required to produce handcrafted features as the inputs of CrackNet. The data depth used at hidden layers of CrackNet is 360, resulting in more than 1 million parameters. As the developing of semantic segmentation technologies, various end-to-end pixel-level detection algorithms have been proposed. Long et al. [11] proposed a symmetrical Full Convolutional Neural Network (FCN). However, the FCN model requires three training sessions to achieve a better training result. In order to solve this problem, U-Net [12] is proposed, which can provide accurate segmentation results after only one training session at small datasets. However, all feature mappings are used to transfer the encoder to decoder in U-Net, and then concatenation is used to convolute in series, which makes a large model framework and requires high memory. Huang et al. [13] introduce the Dense Convolutional Network (DenseNet). Instead of adopting deeper and broader networks, DenseNet greatly reduces the amount of network parameters through reusing the feature maps and connecting each layer to the other layer in a feed-forward fashion. However, since the current deep learning framework does not support DenseNet's dense connections, the output of the previous layer can be stitched together with the output of the current layer by means of repeated concatenation operations and then passed to the next layer. To solve these problems, the SegNet model was proposed [14]. The first 13 convolutional layers of the VGG-16 network [15] are used in the encoder, and the max-pooling indices are stored for being used in the corresponding decoder to perform non-linear upsampling. This eliminates the requirement of learning to upsample. Therefore, in SegNet, the number of trainable parameters is small, and the memory requirement is low. In this work, SegNet model is modified to realize an end-to-end pixel-wise crack detection.

The remainder of this paper is organized as follows. In Sect. 2, the architecture of SegNet and the performance evaluation indicators are introduced. Section 3 details the collection and labelling of crack image datasets. In Sect. 4, the SegNet-based crack detection model is retrained and the performance are evaluated. Finally, some conclusions are given.

2   Architecture of SegNet and the Performance Evaluation Indicators

2.1   Architecture of SegNet

SegNet uses an encoder-decoder structure to achieve end-to-end semantic segmentation. As shown in Fig. 1, the encoder of the SegNet is modified from the VGG-16 and is symmetrical with the decoder. A decoder upsamples the input using the transferred pool indices from the corresponding encoder. The Softmax layer classifies each pixel independently using the features input by the last decoder. The blue portion represents the convolutional layer, the green portion represents the pooled layer, the red portion represents upsampling, and the yellow portion represents the Softmax classifier.

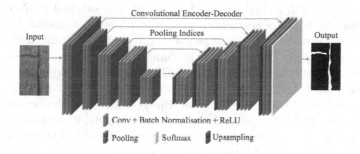

Fig. 1. SegNet architecture. (Color figure online)

SegNet coding method is the same as the CNN convolution operation. Convolution layer is used to extract features, and then pooling operation is used to reduce the size of input feature maps as well as the number of parameters, and therefore speed up the model training. SegNet's pooling layer has an index function compared to pooling layers in the other architectures. It is used to record the location of the maximum pooling results relative to the pooling core during the pooling process. The pooling layer (green) and the upsampling (red) are connected by pool indices, which outputs the pooling indices to the corresponding upsampling. Since the network uses a symmetrical structure, the first pooling layer corresponds to the last upsampling. The SegNet decoding process uses zero-fill convolution, which is mainly used to fill up the upsampled feature map information, so that the information discarded by the pooling operation is obtained during the decoding process. Compared to FCN's transposed convolution strategy, SegNet uses index to directly put the data back to the corresponding position in the upsampling layer, and then performs convolutional layer training. Therefore, in the upsampling process, the training and learning processes are eliminated and only a small memory is occupied to store the index information, which improves the decoding efficiency. SegNet utilizes the largest pooling indices to directly upsample the feature map and convolve it with the trainable decoder filter bank. The FCN generates a decoder output by deconvolving the input feature map that is added by a corresponding encoder feature map.

2.2 Performance Evaluation Indicators

In semantic segmentation, the Jaccard coefficient, i.e. Intersection Over Union (IOU or IU), is often used as the performance evaluation indicator. The higher the value of IOU, the higher the overlap rate between the target area predicted by the model and the actual marked area. When the IOU value is greater than 0.7, it can be considered that there is a good consistency between the predicted result and the true value. When the IOU value is 1, the predicted target area and the actual marked area are completely overlap.

In this paper, the original SegNet is modified to accommodate this experiment. The main modifications are as follows: (i) The original SegNet model is changed from 21 types of segmentation to two types of segmentation, that is, the multi-classification case is changed to the two-classification, and the input is grayscale image; (ii) The input size is changed from the original 360×480 to 256×256; (iii) To prevent over-fitting, a BN layer [16] is added between every convolutional layers and the ReLU activation function; (iv) The Adam optimizer [17] (learning rate 0.0001) is used to be the gradient optimization strategy in order to increase the convergence rate. At the same time, if the learning effect is not improved after 5 epochs, the learning rate is attenuated to one-fifth of the original one, and the lowest learning rate is 0.00001.

3   Crack Image Dataset

3.1   Data Acquisition and Labeling

A crack image dataset is utilized in this work, including 2068 images at the resolutions of 1024×1024 with an image classification label, which is published by Ma Weifei et al. [18]. To be used for training the SegNet, a pixel-wise annotation is required. The annotation tool in this article is a Web-based Semantic Segmentation Editor developed by Hitachi Automotive Industry Lab. It can annotate the ordinary two-dimensional images. It can automatically create a polygon to fit the sharply contrasted edge contours through contrast threshold detection, then the binarized label is created.

In order to highlight the cracks in the image, this annotation uses a binarized black and white image with grayscales 0 and 255. The bridge crack is marked in white and the background is marked in black.

3.2 Data Extension

The bridge crack images at the size of 1024 × 1024 are too large to train SegNet. Therefore, the large image is automatically cut into a serial of small images with the size of 256 × 256 by a code written in Python. After removing the blurred images, 5180 small images were obtained finally. The samples are divided into training set, validation set, and test set in a ratio of 8:1:1.

3.3 Sliding Window Scanning Arbitrary Size Image for Detection

Since the input image size of the training model is 256 × 256, the image size input during prediction must also be 256 × 256. However, in the actual experiment process, the images taken are often at a large size and it is unrealistic to cut the images into 256 × 256 and then detect the cracks in the image. Therefore, this experiment uses sliding window scanning technique to achieve crack detection of large images.

Considering the size of the large images is not exactly a multiple of 256 × 256 pixels, the border zero padding operation is first given to the image. The image is expanded to an integer multiple of 256 in length and width by zero-padding. Then, the filled image is cut from left to right from top to bottom with a size of 256 × 256, and the cut small image is predicted in order and the predicted result that is placed at the corresponding position of the graph A. Finally, the complete prediction graph A is obtained.

4 Experiments

4.1 Training Process

The training parameter configuration of training the SegNet is shown in Table 1, the training process of the model is as follows: (i) The original dataset images and the label images are normalized; (ii) Each forward propagation takes 4 samples from the training set of step (i), and the obtained output value is compared with the labeled sample to get the loss value, through which the model weights are updated by the Adam algorithm; (iii) For each training epoch, the verification set is used to verify the accuracy. A total of 100 epochs are trained on the training set; (iv) Finally, the trained model is used to predict the results of the test set, and the model performance is evaluated by the IOU.

Table 1. Configuration of the model training parameters.

Parameter name	Value
Input image size	256×256
Epoch	100
Batch-size	4
Optimizer	Adam (learning rate 0.0001)
Learning rate attenuation	The lower limit is 0.00001

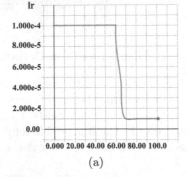 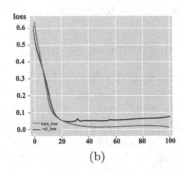

(a)   (b)

Fig. 2. Learning rate and Loss value curve: (a) Learning rate, (b) Loss value.

4.2 Comparative Analysis of Training Results

The SegNet experimental model is trained 100 epochs. Not every parameter is saved during training. If a subsequent training, does not achieve a better loss value than the previous training on the verification set, it will not be saved.

It takes about 17 h to train the model. The following is an analysis of the changes of parameters in the process of model training. As shown in Fig. 2(a), the learning rate decreases to 1/10 at the 68th epoch. At this point, the learning rate decays to the lower limit of the setting. As shown in Fig. 2(b), the loss value on the training set is generally decreasing slowly, but the loss value on validation set does not fall after the 60th epoch. This experiment preserves the best model based on the loss value of the validation set. The final model has an IOU value of 0.782 on the test set.

Figure 3 shows part of the prediction result. Images in the first column, the second column and the third column correspond to the original images, the annotation images and the prediction results. It can be seen from the figure that the model can still extract crack information better for images with relatively complex cracks and strong background interference.

The predicted results of the SegNet model are also compared with the results of Canny and Sobel algorithms, as shown in Fig. 4. The Canny algorithm is easily interfered by rough surface impurities in the bridge crack image. Similarly, the Sobel edge detection algorithm is also difficult to distinguish cracks from stains,

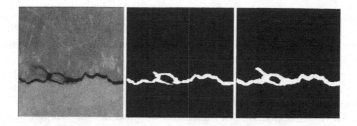

Fig. 3. Partial prediction effect diagram on bridge crack test set.

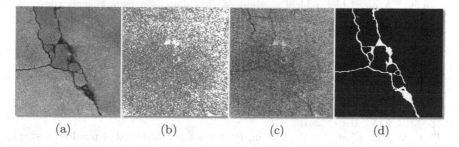

(a)   (b)   (c)   (d)

Fig. 4. Comparison of crack detection results between SegNet and traditional methods: (a) Original image, (b) Canny, (c) Sobel, (d) SegNet.

potholes, and scratches. In contrast, SegNet can accurately extract crack regions from complex background. Therefore, the realized crack detection methodology based on SegNet is effective.

5   Conclusion

In this paper, a crack detection methodology is realized based on SegNet. It is trained by an image dataset with 5180 bridge crack images. Sliding window technique is utilized to ensure the model can cope with the images at larger size. Experiment results show that the IOU on the bridge crack image dataset reaches more than 0.7. Compared with traditional edge detection methodologies, the SegNet method used in this paper can detect cracks in the images with large noise interference and complex background texture.

Acknowledgements. This work is financially supported in parts by the Fujian Provincial Department of Science and Technology of China (Grant No. 2019H0006 and 2018J01774), the National Natural Science Foundation of China (Grant No. 61601127), and the Foundation of Fujian Provincial Department of Industry and Information Technology of China (Grant No. 82318075).

References

1. Schmidt, P., Earthman, J.: Development of a scanning laser crack detection technique for corrosion fatigue testing of fine wire. J. Mater. Res. **10**(2), 372–380 (1995)

2. Tong, C., Wang, W.: Research of detection of a crack on pressure vessel based on both of acoustic emission diagnosis and ultrasonic testing. In: 2011 Second International Conference on Mechanic Automation and Control Engineering, pp. 1510–1513. IEEE (2011)

3. Fu, Z., et al.: Image segmentation with multilevel threshold of gray-level & gradient-magnitude entropy based on genetic algorithm. In: 2015 International Conference on Artificial Intelligence and Industrial Engineering

4. Joseph, H., Periyasamy, R.: An analytical method for the adaptive computation of threshold of gradient modulus in 2D anisotropic diffusion filter. Biocybernetics Biomed. Eng. **37**(1), 1–10 (2017)

5. Zhu, C.: A novel hierarchical method of ship detection from spaceborne optical image based on shape and texture features. IEEE Trans. Geosci. Remote Sens. **48**(9), 3446–3456 (2010)

6. Ciresan, D.C.: Deep, big, simple neural nets for handwritten digit recognition. Neural Comput. **22**(12), 3207–3220 (2010)

7. Hou, L., Samaras, D., Kurc, T.M., Gao, Y., Davis, J.E., Saltz, J.H.: Patch-based convolutional neural network for whole slide tissue image classification. In: Proceedings of the IEEE Conference on Computer Vision and Pattern Recognition, pp. 2424–2433. IEEE (2016)

8. Ji, J., Wu, L., Chen, Z., Yu, J., Lin, P., Cheng, S.: Automated pixel-level surface crack detection using U-Net. In: International Conference on Multi-disciplinary Trends in Artificial Intelligence, pp. 69–78. IEEE (2018)

9. Wang, X., Hu, Z.: Grid-based pavement crack analysis using deep learning. In: 2017 4th International Conference on Transportation Information and Safety (ICTIS), pp. 917–924. IEEE (2017)

10. Zhang, A.: Deep learning-based fully automated pavement crack detection on 3D asphalt surfaces with an improved CrackNet. J. Comput. Civil Eng. **32**(5), 04018041 (2008)

11. Long, J., Shelhamer, E., Darrell, T.: Fully convolutional networks for semantic segmentation. In: Proceedings of the IEEE Conference on Computer Vision and Pattern Recognition, pp. 3431–3440. IEEE (2015)

12. Ronneberger, O., Fischer, P., Brox, T.: U-Net: convolutional networks for biomedical image segmentation. In: International Conference on Medical Image Computing and Computer-Assisted Intervention, pp. 234–241. IEEE (2015)

13. Huang, G., Liu, Z., Van Der Maaten, L., Weinberger, K.Q.: Densely connected convolutional networks. In: Proceedings of the IEEE Conference on Computer Vision and Pattern Recognition, pp. 4700–4708. IEEE (2017)

14. Badrinarayanan, V., Handa, A., Cipolla, R.: SegNet: a deep convolutional encoder-decoder architecture for robust semantic pixel-wise labelling. arXiv preprint arXiv:1505.07293 (2015)

15. Simonyan, K., Zisserman, A.: Very deep convolutional networks for large-scale image recognition. arXiv preprint arXiv:1409.1556 (2014)

16. Ioffe, S., Szegedy, C.: Batch normalization: accelerating deep network training by reducing internal covariate shift. arXiv preprint arXiv:1502.03167 (2015)

17. Kingma, D.P., Ba, J.: Adam: a method for stochastic optimization. arXiv preprint arXiv:1412.6980 (2014)

18. Ma, W.F.: Research on Bridge Crack Detection Algorithm Based on Deep Learning. MS thesis, Shaanxi Normal University (2018). (in Chinese)

Statistical Analysis of the Performance of the State-of-the-Art Methods for Solving TSP Variants

Boldizsár Tüű-Szabó[1]($\boxtimes$), Péter Földesi[2], and László T. Kóczy[1,3]

[1] Department of Information Technology, Széchenyi István University,
Győr, Hungary
{tuu.szabo.boldizsar,koczy}@sze.hu
[2] Department of Logistics, Széchenyi István University, Győr, Hungary
foldesi@sze.hu
[3] Department of Telecommunications and Media Informatics,
Budapest University of Technology and Economics, Budapest, Hungary

Abstract. In this paper we analyze the efficiency of the state-of-the-art methods for solving two TSP variants, the Traveling Salesman Problem with Time Windows and one-commodity Pickup-and-Delivery Traveling Salesman Problem. Three models (polynomial, exponential, square-root exponential) were fitted to the mean run times of each method. The parameters of the curves, the R^2-values and the RMSE values were compared.

Keywords: Traveling Salesman Problem · Time windows · Pick up · Delivery

1 Introduction

The Traveling Salesman Problem with Time Windows (TSPTW) is an extension of the classical TSP. In this problem the salesman has to visit all customers within a specific time window starting his journey from the depot and then returning to the depot. Each customer has a service time and a time window defining its ready time and due time. If the salesman visits a customer after its due time then this tour is called infeasible because it does not satisfy the constraints. If the salesman arrives to a customer before its ready time, he must wait because he cannot leave before it. The task is to find the minimum cost tour visiting each customer, which satisfies the time windows (a feasible tour with minimum cost). The three most efficient methods were chosen for the statistical analysis: the Discrete Bacterial Memetic Evolutionary Algorithm (DBMEA) [7], the general VNS heuristic [2] and the Compressed Annealing [10]. The DBMEA algorithm is a population based evolutionary algorithm which combines the discretized bacterial evolutionary algorithm with 2-opt and 3-opt local search techniques. The general VNS heuristic consists of two parts: constructive and optimization stages. In the first stage, the heuristic constructs a feasible solution using VNS, and in the optimization stage the heuristic improves the feasible solution with a General VNS heuristic. The Compressed Annealing combines the simulated annealing with a variable penalty method.

R. Chamchong and K. W. Wong (Eds.): MIWAI 2019, LNAI 11909, pp. 255–262, 2019.
https://doi.org/10.1007/978-3-030-33709-4_23

The one-commodity Pickup-and-Delivery Traveling Salesman Problem (1-PDTSP) is an extended version of the classical Traveling Salesman Problem. It was introduced by Hernández-Pérez in 2004 [4]. In this problem a set of nodes and the costs between them are given. There is a specific node, the depot which is always the starting point of the tours. One kind of products has to be transported from some nodes to others. The customers are divided into two groups: the delivery (they supply a given amount of product) and pickup customers (they demand a given amount of product). The value of demanding or delivering amount is assigned to each node. The products are transported with one vehicle which has a maximum capacity. The aim is to find a feasible solution with lowest cost which visit each node. In metric case each node is visited only once in the optimal tour. The three most efficient methods were selected from the literature and analyzed: the DBMEA [11], the Mix-GVNS heuristic [9] and the GRASP/VND heuristic [5]. The GRASP/VND is a hybrid heuristic method that combines a greedy randomized adaptive search procedure (GRASP) with variable neighborhood descent (VND). The Mix-GVNS is a variable neighborhood search approach with k-opt, double-bridge and insertion operators.

For the classical TSP some papers can be found in the literature which have been dedicated to statistically investigating the performance of the TSP solver algorithms. Plotting the run time of the tested instances Applegate *et al.* claimed that the Concorde may scale exponentially [1]. Hoos and Stützle analyzed the run time behavior of the Concorde algorithm on random uniform Euclidean (RUE) TSP instances [6]. The root-exponential scaling model was the most accurate for run times over sets of RUE instances. Mu *et al.* examined the empirical performance of EAX and Lin-Kernighan heuristic on RUE instances up to 4500 nodes [8]. The analysis indicated that the scaling for Lin-Kernighan heuristic lies somewhere between the polynomial and the square-root exponential model. In our former work we analyzed the performance of the state-of-the-art methods for VLSI TSP instances up to 20000 nodes [12]. Papers which dealing with the statistical analysis of the performance of the solver methods for TSP variants cannot be found in the literature.

2   Computational Results

Two TSP variants were chosen for the analysis: the TSPTW and the 1-PDTSP. In both cases the following three model were fitted to the mean run times of the solver methods on the tested instances:

- polynomial: $f[a,b](n) = a \cdot n^b$
- exponential: $f[a,b](n) = a \cdot b^n$
- square-root exponential: $f[a,b](n) = a \cdot b^{\sqrt{n}}$

A curve which minimizes the root mean square error (RMSE) was fitted to the mean values of the run times.

2.1   Analysis of the TSPTW

The DBMEA algorithm was tested on instances proposed by Gendreau et al. [3], which contains 140 instances grouped in 28 test cases.

First the algorithms were compared in terms of solution quality. In 24 out of 28 test cases the DBMEA algorithm found the best-known results, in 4 cases the general VNS heuristic found slightly lower values than the DBMEA, while the Compressed Annealing failed to find the best-known solution in 14 cases. In terms of tour qualities the general VNS heuristic is the best, the DBMEA is the second best heuristic [7].

Table 1 shows the parameters of the fitted curves of each algorithm. Surprisingly the Compressed Annealing resulted in the lowest b parameters. It indicates that if the algorithms keep their tendencies on large instances Compressed Annealing may become the fastest algorithm on large instances with hundreds of nodes. Further tests on large instances are needed to confirm this hypothesis.

Table 2 shows the RMSE and R^2-value of the fitting models. Because of the low number of points the R^2-values are very close to 1 in each cases. Based on the RMSE values for the general VNS heuristic and for the DMEA the polynomial, for the Compressed Annealing the square-root exponential model shows the highest accuracy.

Figures 1, 2 and 3 show the curve of the fitting models for the three algorithms.

Table 1. Parameters of the fitting models

		a	b
Compressed Annealing	Polynomial	0.185	0.9474
	Exponential	3.194	1.016
	Square-root exponential	1.246	1.282
General VNS heuristic	Polynomial	0.001216	1.798
	Exponential	0.3791	1.026
	Square-root exponential	0.06556	1.538
DBMEA	Polynomial	6.885e-06	3.12
	Exponential	0.2379	1.04
	Square-root exponential	0.01133	2.007

Table 2. The RMSE and R^2-value of the fitting models

		RMSE	R^2-value
Compressed Annealing	Polynomial	0.608	0.9856
	Exponential	0.4023	0.9937
	Square-root exponential	0.1303	0.9993
General VNS heuristic	Polynomial	0.05104	0.9994
	Exponential	0.2732	0.983
	Square-root exponential	0.1301	0.9962
DBMEA	Polynomial	0.2312	0.9983
	Exponential	0.4322	0.994
	Square-root exponential	0.2556	0.9979

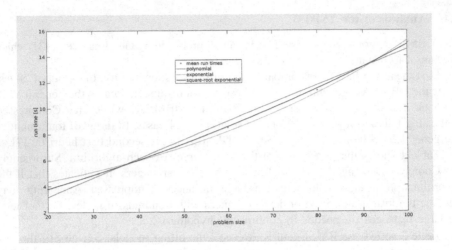

Fig. 1. Fitting models of the Compressed Annealing

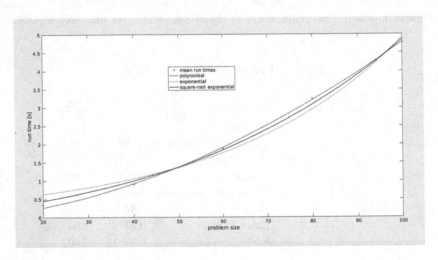

Fig. 2. Fitting models of the general VNS heuristic

2.2   Analysis of the 1-PDTSP

Table 3 contains the parameters of the models for each method. The Mix-GVNS heuristic has the lowest b parameter values. It indicates that it will be the fastest method on large instances, but in terms of tour quality it underperformed compared with the DBMEA and the GRASP/VND method [11].

Table 4 shows the RMSE and R^2-value of each examined fitting model. Due to the low number of testing points the R^2-values are close to 1 so further tests of the algorithms are needed especially on larger instances. For the Mix-GVNS and the

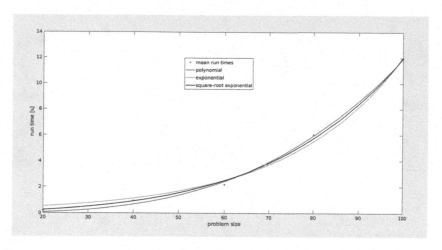

Fig. 3. Fitting models of the DBMEA

Table 3. Parameters of the fitting models

		a	b
Mix-GVNS	Polynomial	1.146e-06	3.659
	Exponential	0.1581	1.051
	Square-root exponential	0.004598	2.352
GRASP/VND	Polynomial	1.073e-07	3.958
	Exponential	0.04253	1.055
	Square-root exponential	0.0009191	2.502
DBMEA	Polynomial	1.195e-06	3.71
	Exponential	0.191	1.052
	Square-root exponential	0.005295	2.385

Table 4. The RMSE and R^2-value of the fitting models

		RMSE	R^2-value
Mix-GVNS	Polynomial	0.1277	0.9998
	Exponential	0.3732	0.9987
	Square-root exponential	0.1886	0.9997
GRASP/VND	Polynomial	0.1056	0.9992
	Exponential	0.0593	0.9998
	Square-root exponential	0.05486	0.9998
DBMEA	Polynomial	0.4096	0.9991
	Exponential	0.6863	0.9975
	Square-root exponential	0.4908	0.9987

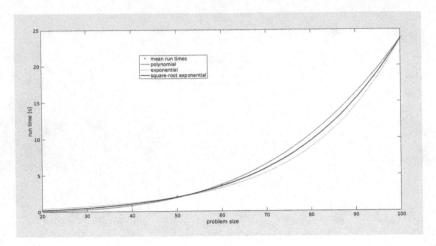

Fig. 4. Fitting models of the Mix-GVNS

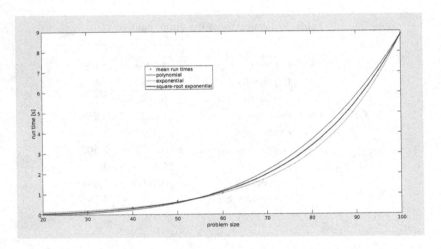

Fig. 5. Fitting models of the GRASP/VND

DBMEA the polynomial, for the GRASP/VND the square-root exponential model resulted in the lowest RMSE values.

Figures 4, 5 and 6 show the curve of the fitting models for the three algorithms.

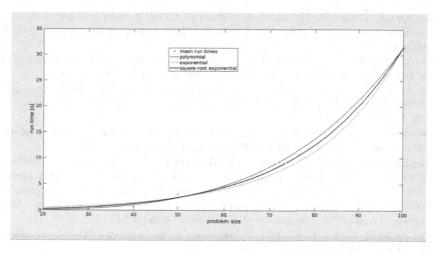

Fig. 6. Fitting models of the DBMEA

3   Conclusion

In this paper two TSP variants, the Traveling Salesman Problem with Time Windows and the one-commodity Pickup-and-Delivery Traveling Salesman Problem were statistically investigated: three fitting models were fitted to the mean run times of the most efficient methods, and the properties of the curves were analyzed.

Acknowledgement. This work was supported by National Research, Development and Innovation Office (NKFIH) K108405, K124055.   Supported by the ÚNKP-18-3 New National Excellence Program of the Ministry of Human Capacities.

References

1. Applegate, D.L., Bixby, R.E., Chvátal, V., Cook, W.J.: The Traveling Salesman Problem: A Computational Study, pp. 489–530. Princeton University Press, Princeton (2006)
2. da Silva, R.F., Urrutia, S.: A General VNS heuristic for the traveling salesman problem with time windows. Discrete Optim. **7**(4), 203–211 (2010)
3. Gendreau, M., Hertz, A., Laporte, G., Stan, M.: A generalized insertion heuristic for the traveling salesman problem with time windows. Oper. Res. **46**(3), 330–335 (1998)
4. Hernández-Pérez, H., Salazar-González, J.J.: Heuristics for the one-commodity pickup-and-delivery travelling salesman problem. Transp. Sci. **38**, 245–255 (2004)
5. Hernández-Pérez, H., Rodríguez-Martín, I., Salazar-González, J.J.: A hybrid GRASP/VND heuristic for the one-commodity pickup-and-delivery travelling salesman problem. Comput. Oper. Res. **36**, 1639–1645 (2009)
6. Hoos, H.H., Stützle, T.: On the empirical scaling of run-time for finding optimal solutions to the travelling salesman problem. Eur. J. Oper. Res. **238**, 87–94 (2014)

7. Kóczy, L.T., Földesi, P., Tüű-Szabó, B.: Enhanced discrete bacterial memetic evolutionary algorithm-an efficacious metaheuristic for the traveling salesman optimization. Inf. Sci. **460**, 389–400 (2017)
8. Mu, Z., Dubois-Lacoste, J., Hoos, H.H., Stützle, T.: On the empirical scaling of running time for finding optimal solutions to the TSP. J. Heuristics **24**(6), 879–898 (2018)
9. Mladenovic, N., Uroševic, D., Hanafi, S., Ilic, A.: A general variable neighborhood search for the one-commodity pickup-and-delivery travelling salesman problem. Eur. J. Oper. Res. **220**, 270–285 (2012)
10. Ohlmann, J.W., Thomas, B.W.: A compressed-annealing heuristic for the traveling salesman problem with time windows. INFORMS J. Comput. **19**(1), 80–90 (2007)
11. Tüű-Szabó, B., Földesi, P., Kóczy, L.T.: The discrete bacterial memetic evolutionary algorithm for solving the one-commodity pickup-and-delivery traveling salesman problem. In: Kóczy, L.T., Medina-Moreno, J., Ramírez-Poussa, E., Šostak, A. (eds.) Computational Intelligence and Mathematics for Tackling Complex Problems. SCI, vol. 819, pp. 15–22. Springer, Cham (2020). https://doi.org/10.1007/978-3-030-16024-1_3
12. Tüű-Szabó, B., Földesi, P., Kóczy, L.T.: Analyzing the performance of TSP solver methods. In: ESCIM 2019, Toledo, Spain (2019)

Identification of Conversational Intent Pattern Using Pattern-Growth Technique for Academic Chatbot

Suraya Alias[(⊠)], Mohd Shamrie Sainin, Tan Soo Fun, and Norhayati Daut

Knowledge Technology Research Unit, Faculty of Computing and Informatics, Universiti Malaysia Sabah, Kota Kinabalu, Malaysia
suealias@ums.edu.my

Abstract. This paper describes the development work of our Academic Chatbot model in identifying the user's conversational intents patterns. We experimented using social conversation log data from WhatsApp Messenger by the academic coordinator and students during the student's internship period. The discovered conversational patterns are used as a heuristic in building the knowledge base for the intent and entity of our Academic Chatbot model. Our preliminary findings depicted that related conversational intent patterns named Frequent Intent Pattern (FIP) was discovered with confidence value as high as 0.9 using the Sequential Pattern-Growth technique. The basis of using a Pattern-Growth pattern representation has given an insight where the chatbot can learn over time and new information can be added based on the intent pattern discovery. The outcome of this project is a customized Academic Chatbot (AcaBot) model that will be able to assist academicians and students in Academic Institution automatically and instantly 24/7 regarding relevant academic topics.

Keywords: Academic chatbot · Sequential pattern mining · Pattern-growth

1 Introduction

In the edge of where information is disseminated without time constraint and boundaries, an automated conversational agent using Natural Language Processing (NLP) and Artificial Intelligence (AI) also known as chatbot has a promising future to assist in improving student interaction in an Academic Institution. The use of chat messaging platform such as WhatsApp, Facebook and Telegram have provided a medium for students and academician to communicate and disseminate information. However, the issue arises when: (1) Some user queries need prompt feedback that is constrained by time, location and availability of the responder. (2) Redundant and similar academic queries from the students make the manual response procedure less effective and time-consuming for the academicians to entertain. (3) Academic-related information is difficult to disseminate in an academic website due to poor website navigation design. Due to this, a new paradigm-shift on students experience in

© Springer Nature Switzerland AG 2019
R. Chamchong and K. W. Wong (Eds.): MIWAI 2019, LNAI 11909, pp. 263–270, 2019.
https://doi.org/10.1007/978-3-030-33709-4_24

communication and information dissemination in an Academic Institution using a chatbot is proposed.

The proposed Academic Chatbot (AcaBot) model tries to combine the best of both chatbot types (structured and AI-driven) to make the chatbot learns over time based on the textual conversation intent pattern discovery. First, the concept of Frequent Text Pattern Mining is extended to identify the conversational intent patterns using the Sequential Pattern-Growth technique. Next, a new Academic ChatBot model is developed on top of the discovered textual conversational intent patterns for pattern matching based on the context's similarity related to Academic topics in order to learn and converse the textual answer. The outcome of this project is customized AcaBot model that will be able to assist academicians and students in Academic Institution automatically and instantly 24/7 regarding relevant Academic topics.

2   Related Works

A chatbot agent can converse via voice and textual method has seen much advancement since the evolution of the proto-chatbot ELIZA by [1]. To date, a chatbot has been pervasively adopted for the use of customer service advisor, health advisor, e-commerce shopping assistant, and in a smart home environment [2]. The adoption of chatbot in an Academic Institution has seen a rising trend where it has been implemented as automated library assistance [3], student academic advisory [4, 5], automated response to a university FAQ [6] and also to assist student's university life [7].

In general, there are two types of chatbot: (1) structured-based and (2) AI-driven [8, 9]. The structured-based relies on the amount of information pumped into the knowledge base and will respond to options selected by the user [5]. For example, an online shopping Lazada chatbot (LazBot) will return the relevant information regarding some product based on the selected button in the menu. In contrast, an intelligent Chatbot with the implementation of AI and NLP can converse naturally with users by understanding and learning the set of questions and providing personalized solutions such as the IBM Watson, Apple Siri, Google Assistant and Amazon Alexa. This AI-driven chatbot implements technique such as AI Markup Language [9], Neural Network-based [10, 11], Sequence to Sequence (Seq 2Seq) model [12], linguistics rule based [13] and also deep learning [14] to better understand human intents and to mimic human conversation. However, most of the aforementioned AI techniques require deep NLP, very high computation algorithm and training data to understand user's intent and providing a valid response to the user which becomes a huge challenge.

Thus, this study proposes an unsupervised approach using text pattern discovery by mining social conversation log data.

Sequential Pattern Mining

The Sequential Pattern Mining (SPM) algorithm can be divided into Apriori-based and Pattern-Growth approach [15]. The main difference is the Pattern-Growth approach implements the "divide-and-conquer" strategy, where else the prior implements the "generate-and-test" strategy.

The use of support and confidence metrics in SPM reflect the significance relationship and strength of the rules discovered from the frequent patterns. In Text Processing, pattern in a text can be represented using frequent sequences where each sentence is treated as a sequence. For instance, the sentence "LI logbook submission" is a sequence that consists of terms <{LI}{logbook}{submission}>. Given user specified minimum support threshold min_sup denotes as σ, a pattern P is said as frequent or Frequent Pattern (FP), if $sup(P) \geq \sigma$. If the pattern is an order-respected, it is known as Frequent Sequential Pattern (FSP) which is useful in mining text data where the order of sequence words is crucial to the content discourse.

The next section presents the proposed AcaBot framework that implements the Sequential Pattern-Growth technique in the Intent Identification phase.

3 The AcaBot Framework

Definition
Utterance: Sentence that being uttered or conversed by the user either by text or speech. For instance, if a user types a message line "I want to ask about the LI report submission.", the whole message line is considered as the utterance.
Intent: An intent is the user's intention in a conversation. Using the above example, the user's intent is to ask regarding a report submission. Usually, an intent name is set using a verb and noun convention such as "submitReport".
Entity: An entity can modify an intent to be more specific and descriptive. Such as the entity found based on the utterance example above is "LI" that stands for *Latihan Industri* or internship.

The AcaBot model consists of three main layers:

1. AcaBot Knowledge Base,
2. AcaBot Intent Identification, and
3. AcaBot Conversation Engine.

Figure 1 illustrates the AcaBot model framework, where the focus of this paper is on the automatic intent identification discovery via Pattern-Growth technique for the development of the AcaBot model knowledge base. The process starts when user enters a text message to start the conversation. At the same time, Acabot will prompt a menu-structured options to assist the academic-related conversation process. From the conversation, the Acabot Intent Identification process will be performed. The next process is to match and rank the text features in order to classify user's intent or queries by discovering the text pattern. If the text matches any of the classes, Acabot will perform a lookup in the Acabot Knowledgebase to prepare the respond to the user via Acabot Conversation Engine. Else, if no matches, options to escalate the queries or autosuggestion to related information will be prompted. This process will iterate until user ends the conversation.

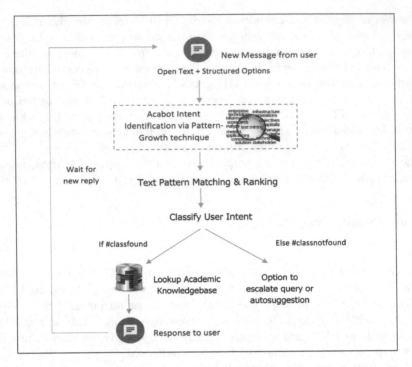

Fig. 1. The AcaBot model framework

3.1 AcaBot Knowledge Base

To develop the AcaBot Knowledge Base (KB), we seek cooperation from the intern-ship coordinator to provide a sample of WhatsApp messages logs with their students before and during the internship period. By mining the text conversation data, our goal is to identify and discover textual conversational patterns in order to design the sets of intents, entity and response related to academic topics to the users.

The chatbot dataset is given in Table 1. The conversation data text includes English and Malay words, emoticons and short forms.

Table 1. Details of conversation dataset for AcaBot knowledge base.

Dataset item	Details
Total WhatsApp conversation files	12
Conversation period	05/10/2017–18/07/2018
Conversation type	Group chat (1), personal chat (11)
Conversation topic	*Latihan Industri (LI)*/internship
Total conversation lines	537
Total terms (with stop words applied)	1834
Total terms (without stop words applied)	1977
Total malay stop words	334
Total English stop words	250

Sample of conversation text excerpt is illustrated in Fig. 2 (the name is anonymous for confidentially). From the conversation, the student's intention is regarding the submission of the *Latihan Industri (LI)* forms, reports and logbook. For this type of frequent repetitive questions, rather than manually setting the intent and entity using on-the-shelf or an opensource chatbot framework that is available, the proposed AcaBot can learn over time based on the automated intent patterns discovery from the conversation data. Thus, this can benefit the AcaBot implementation, where new information can be added based on the intent discovery.

```
16/06/2018, 3:57 pm - Student: Hi, Dr. Regarding the LI-4B form, I need
to submit the form on smart2 if LI SV sent to academic SV?
16/06/2018, 5:40 pm - Student: Hi, Dr. Regarding the first submission of
LI report, the deadline is one month before the internship end or must
submit on 18/6?
16/06/2018, 6:15 pm - Student: Hi, Dr. Regarding the logbook, must sub-
mit the logbook that ald view and sign by li sv or just submit every-
thing we wrote?
16/06/2018, 8:46 pm - Internship Coordinator: Hi. Submit everything in
smart2 via the submission link.
18/06/2018, 5:50 pm - Student: Sorry, Dr. I want to ask again about the
submission.
18/06/2018, 5:51 pm - Student: We only able to upload one file, means we
need to zip the three files. But my li sv hvnt give back the logbook
18/06/2018, 5:52 pm - Student: So I need to wait for the logbook only
submit?
18/06/2018, 5:52 pm - Internship Coordinator: Yes
```

Fig. 2. Sample of conversation excerpt between student and internship coordinator.

Preprocessing

The text pre-processing tasks are as follows: For each chat messaging files:

(1) Convert terms to lowercase, (2) Tokenize each term using space delimiters, (3) Remove punctuations, numbers and emoticons. (4) Remove Malay and English stop words, (5) Split each conversation line into a list of sentences.

3.2   AcaBot Intent Identification

In this preliminary stage, a Pattern-Growth text representation named Frequent Intent Pattern (FIP) is developed based on the prior work of Frequent Adjacent Sequential Patterns (FASP) by [16] to represent the conversation data for the AcaBot model. The essence of the FASP representation is based on "divide and conquer" approach to extract frequent textual patterns found in adjacent from text data. The FASP representation had been tested on Malay and English news dataset in the area of Text Summarization. Following that, since the FASP representation is data-driven and non-language dependent, the application to unstructured social dataset has given a new insight into the intent and entity identification procedure.

Step 1: Initiate prefixIntent list

Firstly, a set of frequent terms of 1-sequences was set as the *prefixIntent* list denoted as α. Here, each line by line messages is treated as a conversation sentence and stored in an ordered sequence list $cs = \{t_1, t_2, t_k..t_m\}$. For each conversation sentence cs, the utterance support denoted as $uSupp(P)$ for each term t found in the inspected conversation sentence line n is calculated. The sum of $uSupp(P)$ is used as weight w for each t in the feature-vector implementation.

Step 2: FIP Generation

After the *prefixIntent* initialization, we divide the search space of each conversation sentence based on the *prefixIntent* list. Each term in the *prefixIntent* list is used as the *prefix* to recursively conquer the cs list. This is done by joining the α initialized earlier with the next *adjacent sequence* term $t_{(k+1)}$ denoted β found in each cs. The term constraint is applied here where $\beta \in \alpha$ to reduce the effort of generating non-frequent sequences. This constraint is useful as a filter of generating less useful intent pattern found in the conversation. The generation of non-redundant FIP until the term vector size of m follows the Eq. 1:

$$FIP_m = \alpha \cup \beta, \quad \text{based on } cs \tag{1}$$

Step 3: FIP Rules Discovery

This study refers to the SPM Sequential Rule defined in Eq. 2:

$$
\begin{aligned}
&X \rightarrow Y \text{ is a Sequential Rule if;} \\
&conf(X \rightarrow Y) \geq \min\_conf; \ where \\
&support(X \rightarrow Y) = support(X \cup Y) \\
&and \ conf(X \rightarrow Y) = support(X \cup Y)/support(X)
\end{aligned}
\tag{2}
$$

4 Preliminary Findings

In this experiment, the text conversation dataset consists of 12 WhatsApp files (1 chat group, 11 personal senders) with the number of conversation lines is 537. Our preliminary findings resulted that the number of Frequent Intent Pattern (FIP) from 1 until 3-sequences discovered is 619 when using the $min\_sup \geq 2$, meaning that the pattern occurs or uttered more than 2 times in the text conversation lines. Table 2 depicts some output of FIP discovered, written in the form of "<FIP > :$uSupp(P)$".

Table 2. Sample frequent intent pattern discovery

Length	Frequent intent patterns with support > 2
1-sequences	<{li, 347},{industrial,11}, {training,10}{briefing,12},{forms,13}>
2- sequences	<{industrial}{training},10>, <{li}{briefing},7>, <{li}{forms},4>
3-sequences	<{industrial}{training}{briefing},3>

Based on Table 3, let X represents the term *"industrial"* which has the support of 11. This means the term has been uttered or occurred 11 times in the conversation sentence (regardless of the sender). The support for the FIP of 2-sequences *"industrial →* *training"* represented by Y is 10, meaning that it occurred 10 times from the set. From here, we can generate a rule such as, if given the *prefixIntent* as *"industrial"*; then, it was found that there is 0.9 confidence value *Conf* that the term *"training"* can occur together in adjacent to represent user's intent in the respective conversation.

Table 3. Sample of FIP rules discovery

#	FIP (X → Y)	Support	Confidence
1	Industrial	11	–
2	Industrial → training	10	0.9
3	Industrial training → briefing	3	0.3
4	Industrial training → visit	5	0.5
5	Industrial training → evaluation	4	0.4

Another example for the FIP of 3-sequences is "industrial training visit" with the support of 5. Let X now represents FIP of "industrial training" and Y represents "industrial training visit". The Conf value of the rules "industrial training" → "industrial training visit" was found at 0.5. This indicates that if the term "industrial" and "training" occurred together, then there is 0.5 Conf probability that the term "visit" will also occur (in adjacent sentence conversation order).

Based on the preliminary findings, from a user's conversation or utterance, the study is motivated to discover their intent patterns automatically using the FIP. The intent patterns can be used later as features to build the AcaBot knowledge base and to design the dialog flow of response to the user. For example, the sample intent can be something related to "infoIndustrialTraining", while the related entities are "briefing", "visit" and "evaluation" based on the preliminary findings depicted earlier.

5   Conclusion and Future Works

In this paper, we have presented our preliminary findings on user's intent discovery from conversation dataset in order to develop our Academic Chatbot. The basis of using a Pattern-Growth representation has given an insight where the chatbot can learn over time and new information can be added based on the unsupervised intent discovery. Next, our plan is to proceed to create an automated dialog flow module for the conversation engine based on the findings for our AcaBot prototype.

References

1. Weizenbaum, J.: ELIZA - computer program for the study of natural language communication between man and machine. Commun. ACM **9**(1), 36–45 (1966)
2. Pereira, J., Díaz, Ó.: Chatbot dimensions that matter: lessons from the trenches. In: Mikkonen, T., Klamma, R., Hernández, J. (eds.) ICWE 2018. LNCS, vol. 10845, pp. 129–135. Springer, Cham (2018). https://doi.org/10.1007/978-3-319-91662-0_9
3. Meincke, D.: Experiences Building, Training, and Deploying a Chatbot in an Academic Library (2018)
4. Nwankwo, W.: Interactive advising with bots: improving academic excellence in educational establishments. Am. J. Oper. Manag. Inf. Syst. **3**(1), 6 (2018)
5. Ghose, S., Barua, J.J.: Toward the implementation of a topic specific dialogue based natural language chatbot as an undergraduate advisor. In: International Conference on Informatics, Electronics & Vision (ICIEV). IEEE (2013)
6. Ranoliya, B.R., Raghuwanshi, N., Singh, S.: Chatbot for university related FAQs. In: 2017 International Conference on Advances in Computing, Communications and Informatics (ICACCI). IEEE (2017)
7. Dibitonto, M., Leszczynska, K., Tazzi, F., Medaglia, C.M.: Chatbot in a campus environment: design of LiSA, a virtual assistant to help students in their university life. In: Kurosu, M. (ed.) HCI 2018. LNCS, vol. 10903, pp. 103–116. Springer, Cham (2018). https://doi.org/10.1007/978-3-319-91250-9_9
8. Hardalov, M., Koychev, I., Nakov, P.: Towards automated customer support. In: Agre, G., van Genabith, J., Declerck, T. (eds.) AIMSA 2018. LNCS (LNAI), vol. 11089, pp. 48–59. Springer, Cham (2018). https://doi.org/10.1007/978-3-319-99344-7_5
9. Abdul-Kader, S.A., Woods, J.: Survey on chatbot design techniques in speech conversation systems. Int. J. Adv. Comput. Sci. Appl. **6**(7) (2015)
10. Wu, J., et al.: NADiA-towards neural network driven virtual human conversation agents. In: Proceedings of the 17th International Conference on Autonomous Agents and MultiAgent Systems. International Foundation for Autonomous Agents and Multiagent Systems (2018)
11. Gao, J., Galley, M., Li, L.: Neural approaches to conversational AI. In: The 41st International ACM SIGIR Conference on Research & Development in Information Retrieval. ACM (2018)
12. Qiu, M., et al.: Alime chat: a sequence to sequence and rerank based chatbot engine. In: Proceedings of the 55th Annual Meeting of the Association for Computational Linguistics (vol. 2: Short Papers) (2017)
13. Saini, A., Verma, A., Arora, A., Gupta, C.: Linguistic rule-based ontology-driven chatbot system. In: Bhatia, S.K., Tiwari, S., Mishra, K.K., Trivedi, M.C. (eds.) Advances in Computer Communication and Computational Sciences. AISC, vol. 760, pp. 47–57. Springer, Singapore (2019). https://doi.org/10.1007/978-981-13-0344-9_4
14. Yan, R.: "Chitty-chitty-chat bot": deep learning for conversational AI. In: IJCAI (2018)
15. Adamo, J.-M.: Data Mining for Association Rules and Sequential Patterns: Sequential and Parallel Algorithms. Springer, New York (2012). https://doi.org/10.1007/978-1-4613-0085-4
16. Alias, S., et al.: A text representation model using sequential pattern-growth method. Pattern Anal. Appl. **21**(1), 233–247 (2018)

Road Sign Detection and Recognition of Thai Traffic Based on YOLOv3

Paitoon Thipsanthia$^{(\boxtimes)}$ ⓘ, Rapeeporn Chamchong ⓘ,
and Panida Songram ⓘ

POLAR Lab, Faculty of Informatics, Mahasarakham University,
Maha Sarakham, Thailand
{paitoon.thi, rapeeporn.c, panida.s}@msu.ac.th

Abstract. This paper aims to apply a YOLOv3 technique for detecting and recognizing Thai traffic signs in real-time environments. The Thai Traffic Sign Dataset (TTSD) was collected by car cameras to store the video images using the resolution of 1920 × 1080 pixels using 60 frames per second, and a 1280 × 720 pixels and 30 frames per second. In addition, the data was collected in the rural area of Maha Sarakham Province and Kalasin Province. The dataset was generated and distributed for general traffic sign detection and recognition. Two architectures (YOLOv3 and YOLOv3 Tiny) are compared with 50 classes of road signs and 200 badges in each class, containing 9,357 images. The experiment shows that the mean average precision (mAP) of YOLOv3 (88.10%) is better than YOLOv3 Tiny (80.84%) while the speed of YOLOv3 marginally is better than YOLOv3.

Keywords: Road sign detection · Deep learning · Machine learning

1 Introduction

The development of autonomous car technology has gained popularity in several developed countries with a number of reasons. There are many road traffic accidents resulting in serious injury or death because of failure to read road signs [1]. Nowadays, advance computer technology can be proceeded large amounts of data with real time processing. However, the recognition rate of road sign software is not promising. Although some of the most popular works have a high recognition rate, the processing cannot be done quickly enough [2].

Real-time traffic detection and recognition is one of the systems for driving unmanned vehicles that still has problems in time complexity and low accuracy. Yuan et al. [2] proposed the real-time traffic detection and recognition techniques using traffic detection techniques with Aggregated Channel method Features (ACF). This method has high traffic recognition efficiency but low detection rate. Time complexity of traffic sign detection is an average of 13.16 frames per second using CPU processing. It is not fast enough to be used in real-time processing but it provides good recognition performance. This technique applied the Support Vector Machines (SVM) to recognize traffic signs and achieve at 97.78%.

© Springer Nature Switzerland AG 2019
R. Chamchong and K. W. Wong (Eds.): MIWAI 2019, LNAI 11909, pp. 271–279, 2019.
https://doi.org/10.1007/978-3-030-33709-4_25

At present, traffic sign detection and recognition systems can be done faster in everyday life, it also comes with a high cost due to the problems of high speed processing. There are many techniques such as the Mask Regions with Convolutional Neural Network (Mask R-CNN) [3], Single Short Multibox Detector (SSD) [4] and You Only Look Once (YOLO) techniques [5]. These techniques provide high processing speed and efficient recognition so these techniques therefore are popular and wide uses. However, they still cannot classify the similarity object that has different class. Only the research of Zhang et al. [6], who introduced the YOLOv2 technique, categorized the labels according to the types which were divided into three types, forced labels, warning signs and banned signs. As these types have different shape characteristics, YOLO technique can classify with high accuracy. If the traffic signs such as the sign "forced to turn left" and the sign "forced to turn right" have the same size, similar shapes and symbols. It will be difficult to identify traffic sign types. This problem leads to challenge research in detection and recognition of traffic signs in real time for higher efficiency.

The proposal of this paper is road sign detection and recognition of Thai traffic based on YOLOv3. The Thai Traffic Sign Dataset (TTSD) has been collected and distributed for future work. The YOLOv3 and YOLOv3 Tiny has been compared in this study for applying the real time detection and recognition. The paper is organized as follows. In the next section, related works are described. Then, Sect. 3 explains the modification network architecture of YOLOv3. Section 4 presents the data collections and ground truth generation follows by experimental results in Sect. 5. The final section is a conclusion.

2 Related Work

Recently, Convolutional Neural Networks (CNN) is one of deep learning research. CNN has been used to extract and classify objects with promising results. One of the most common problems is how to design a neural network structure. In practice, the well-known structures of ImageNet have been widely used in these research areas. These tasks were developed by many researchers as follows.

The first important task is a selective search to determine an area of an object instead of checking every pixel of an image. This work also offers an object classification system using the Scale Invariant Feature Transform (SIFT) [7] that is calculated from many color spaces. This feature extraction is called handcrafted features. The SVM is used to identify the selected objects. Girshick et al. [8] developed the Regions-based Convolutional Neural Network (R-CNN) by finding region proposals using a selective search then extracting features using the CNN instead of handcrafted features, follows by classification using SVM. Later, Grishick proposed the Fast R-CNN [9] using a selective search, and deep convolution networks to extract features, classify class and predict bounding boxes. Then, the Faster R-CNN [10] was applied by Ren et al. from the fast R-CNN. This technique composes of three parts; (1) feature extraction using CNN without selective search, (2) the Region Proposal Network (RPN) to determine the proposed objects, and (3) the classification of the proposed objects from RPN and the regression of bounding boxes. Although the Faster R-CNN is very accurate, the problem with the R-CNN family is the speed. Another improved technique which is called YOLO. This technique can perform at the real time processing. YOLO will determine boxes and classes at once with a single stage of network. This object detector is significantly faster than the Faster R-CNN, obtaining 45 FPS on a GPU. Another structure can be seen as the extension of YOLO is the SSD. The SSD proposed to predict various boxes from many scales instead of a single scale like YOLO. This helps to solve any image problems with different scales. SSD only uses upper layer (low resolution) for detection because the bottom layer have high resolution. This therefore performs much worse for small objects. An interesting structure is the Feature Pyramid Network (FPN) developed by Lin et al. [11]. FPN applied the concept of the Faster R-CNN. Pyramid concept was used from bottom layer to upper layer in order to combines low-resolution, semantically strong features with high-resolution. FPN can extract data with high semantic content from every resolution. This principle has also been applied in YOLOv3.

From the study of real-time object detection and recognition problems that are state-of-the-art, for example, YOLOv2, SSD321, R-FCN, SSD513, DSSD513 and FPN FRCN found that the YOLO algorithm works the fastest. YOLOv3 has been used in this study for detecting and recognizing traffic signs in real-time.

You Only Look Once (YOLO) Algorithm
YOLO is a state-of-the-art, real-time object detection system. On a Pascal Titan X, it processes images at 30 FPS and has an mAP of 57.9% on COCO test-dev [12]. YOLOv3 is extremely fast and accurate in mAP measured at 0.5 IOU. YOLOv3 is on the same level as Focal Loss [13] but it is about 4x faster. Moreover, you can easily tradeoff between speed and accuracy simply by changing the size of the model, no retraining required.

Loss Function
The loss ($\mathcal{L}$) [5] consists of two parts, the localization loss ($\mathcal{L}_{loc}$) for bounding box offset prediction and the classification loss ($\mathcal{L}_{cls}$) for conditional class probabilities. Both parts are computed as the sum of squared errors. Two scale parameters are used to control how much we want to increase the loss from bounding box coordinate predictions (λ_{coord}) and how much we want to decrease the loss of confidence score predictions for boxes without objects (λ_{noobj}). Down-weighting the loss contributed by background boxes is important as most of the bounding boxes involve no instance. In the paper, the model sets $\lambda_{coord} = 5$ and $\lambda_{noobj} = 0.5$. The loss is calculated as shown in the equation below.

$$\mathcal{L} = \mathcal{L}_{loc} + \mathcal{L}_{cls} \tag{1}$$

$$\mathcal{L}_{loc} = \lambda_{coord} \sum_{i=0}^{S^2} \sum_{j=0}^{B} 1_{ij}^{obj}[(x_i - \hat{x}_i)^2 + (y_i - \hat{y}_i)^2 + \left(\sqrt{w_i} - \sqrt{\hat{w}_i}\right)^2 + \left(\sqrt{h_i} - \sqrt{\hat{h}_i}\right)^2] \tag{2}$$

$$\mathcal{L}_{cls} = \sum_{i=0}^{S^2} \sum_{j=0}^{B} \left(1_{ij}^{obj} + \lambda_{coord}\left(1 - 1_{ij}^{obj}\right)\right)(c_i - \hat{c}_i)^2 + \sum_{i=0}^{S^2} \sum_{c \in C} (1_{ij}^{obj}(p_i(c) - \hat{p}_i(c))^2 \tag{3}$$

where,

- 1_i^{obj}: An indicator function of whether the cell i contains an object.
- 1_{ij}^{obj}: It indicates whether the j-th bounding box of the cell i is "responsible" for the object prediction (see Fig. 1).
- c_i: The confidence score of cell i, Pr(containing an object) * IoU(pred, truth).
- $\hat{c}_i$: The predicted confidence score.
- C: The set of all classes.
- $p_i(c)$: The conditional probability of whether cell i contains an object of class $c \in C$.
- $\hat{p}_i(c)$: The predicted conditional class probability.

The loss function only penalizes classification error if an object is present in that grid cell, $1_{ij}^{obj} = 1$. It also only penalizes bounding box coordinate error if that predictor is "responsible" for the ground truth box, $1_{ij}^{obj} = 1$.

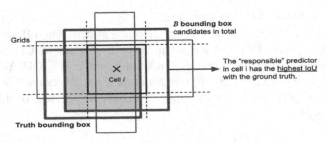

Fig. 1. At one location, in cell i, the model proposes B bounding box candidates and the one that has highest overlap with the ground truth is the "responsible" predictor.

Network Architecture of YOLO

The base model is similar to GoogLeNet with inception module replaced by 1×1 and 3×3 convolution layers. The final prediction of shape $S \times S \times (5B + K)$ is produced by two fully connected layers over the whole convolution feature map (Fig. 2).

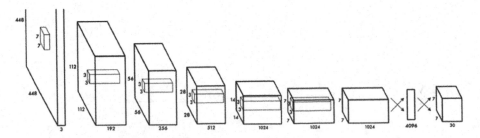

Fig. 2. The network architecture of YOLO [5].

3 The Modification Network Architecture of YOLOv3

The network structures based on YOLOv3 has been applied in this study as follows:

Determination of Neural Network Structures in YOLOv3
The network structure of this study consists of 106 neural networks. The input image size is adjusted using 416×416 pixels, batch size is 24, and learning rate is 0.001. The Darknet53.conv.74 are used for the training set as shown in Fig. 3(A).

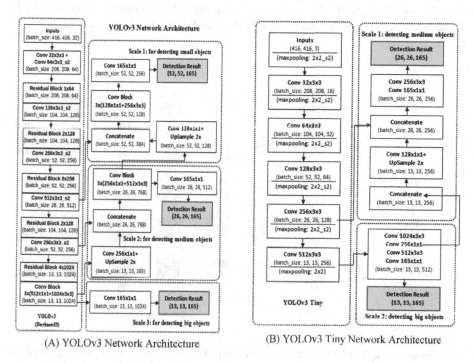

(A) YOLOv3 Network Architecture

(B) YOLOv3 Tiny Network Architecture

Fig. 3. YOLOv3 and YOLOv3 Tiny network architecture.

Determination of Neural Network Structures in YOLOv3 Tiny
The YOLOv3 Tiny has also been applied to compare to YOLOv3. The network structure consists of 23 artificial neural networks. The input image size is adjusted using 416×416 pixels, batch size is 24, and learning rate is 0.001. The Darknet53.conv.74 are used for this study as shown in Fig. 3(B).

4 Data Collections

The collected data was recorded from the car camera by storing video images with the resolution of 1920×1080 pixels using 60 frames per second, and a 1280×720 pixels using 30 frames per second. The recording was collected from 08:00am to

05:00pm on the route from Pathumrat District, Roi Et Province to Yang Talat District, Kalasin Province, and within the area of Mahasarakham University, Khamriang Sub-district, Kantharawichai District of Mahasarakham Province. The driving speed during video recording is between 50 and 70 km per hour, and it records in two minutes of video per a clip.

Characteristics of Collected Data: The dataset was collected as a video image, which was separated into frame by frame. The image frame of traffic signs was selected as a training set and test set. The total classes of signs is at 50. There are 200 badges in each class. The total badges of all classes is 10,000 from 9,357 images. The training set is 90% and test set is 10% by randomizing. The label of each image class is shown in Fig. 4.

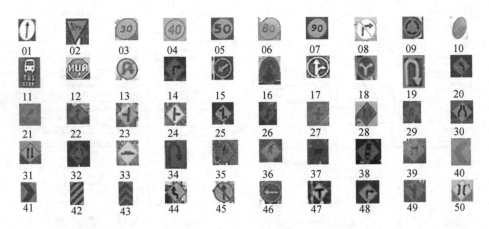

Fig. 4. The 50 classes of the Thai traffic sign dataset.

The Ground Truth of Dataset: The program BBox-Label-Tool [14] was used to determine the actual area of traffic signs that need to detect. The ground truth is stored in a text file (.txt) as shown in Table 1.

Table 1. The ground truth of traffic signs.

No.	File name	Class name	x1	y1	x2	y2
1	20181220-19-00744.txt	01-One_way	376	284	395	299
2	20181220-19-00747.txt	01-One_way	348	285	370	303
⋮	⋮	⋮	⋮	⋮	⋮	⋮
9,357	20190109-183-06279.txt	50-Narrow_ bridge	348	263	379	287

5   Experimental Results

The experiment was run on GPU Nvidia GeForce GTX 1060 with 6 Gb on Ubuntu 16.04 LTS. Fifty classes from 991 files were tested and the evaluation results of each detection and recognition are presented as below.

Thai Traffic Sign Detection Efficiency of YOLOv3

The result of testing was found that the detection and recognition of traffic signs with YOLOv3 obtained mAP = 88.10% (shown in Fig. 5(A)). The average time processing of an image per second is 0.0926, representing an average rate of 10.80 frames per second. This preview real-time processing is at http://tiny.cc/1e5c6y.

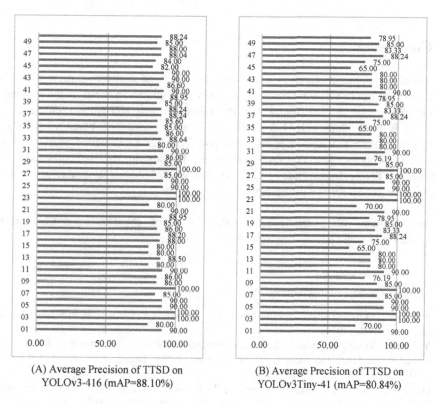

(A) Average Precision of TTSD on
YOLOv3-416 (mAP=88.10%)

(B) Average Precision of TTSD on
YOLOv3Tiny-41 (mAP=80.84%)

Fig. 5. A result of Thai traffic sign detection with YOLOv3 and YOLOv3 Tiny.

Thai Traffic Sign Detection Efficiency of YOLOv3 Tiny

The result of testing was found that the detection and recognition of traffic signs with YOLOv3 tiny obtained mAP = 80.84% (shown in Fig. 5(B)). The average time processing of an image per second is 0.0478, representing an average rate of 20.92 frames per second. This preview real-time processing is at http://tiny.cc/nn5c6y.

A comparison of the results of the Thai traffic sign detection and recognition with YOLOv3 and YOLOv3 Tiny are shown in Table 2. It is found that the detection and recognition with YOLOv3 is better performance than the YOLOv3 Tiny, while speed of YOLOv3 tiny is better than YOLOv3.

Table 2. The results of the TTSD with YOLOv3 and YOLOv3 Tiny.

No.	Model	mAP	Processing time per images
1	YOLOv3	88.10%	10.80 fps
2	YOLOv3Tiny	80.84%	20.92 fps

6 Conclusion

The objective of this research is to develop the detection and recognition with Thai traffic sign dataset in real time based on YOLOv3 technique. In terms of the efficiency of the experiment, it was found that the detection and recognition efficiency of YOLOv3 is better than YOLOv3 Tiny. However, the speed of YOLOv3 Tiny can process faster than YOLOv3. Because YOLOv3 Tiny has a number of neural network layers (23 layers) less than YOLOv3 (106 layers). In addition, we have collected a new of Thailand Traffic Signs Dataset (TTSD) that was published at https://gitlab.com/bombomstory/ttsdb for the future work.

Acknowledgement. We gratefully acknowledge Asst. Prof. Dr. Thawatchai Chomsiri for supporting the PC and GPU for the experiment.

References

1. Sivak, M.: Mortality from road crashes in 193 countries: a comparison with other leading causes of death (2014)
2. Yuan, Y., Xiong, Z., Wang, Q.: An incremental framework for video-based traffic sign detection, tracking, and recognition. IEEE Trans. Intell. Transp. Syst. **18**(7), 1918–1929 (2017)
3. He, K., Gkioxari, G., Dollár, P., Girshick, R.: Mask R-CNN. In: 2017 IEEE International Conference on Computer Vision (ICCV), pp. 2980–2988. IEEE (2017)
4. Liu, W., et al.: SSD: single shot multibox detector. In: Leibe, B., Matas, J., Sebe, N., Welling, M. (eds.) ECCV 2016. LNCS, vol. 9905, pp. 21–37. Springer, Cham (2016). https://doi.org/10.1007/978-3-319-46448-0_2
5. Redmon, J., Divvala, S., Girshick, R., Farhadi, A.: You only look once: unified, real-time object detection. In: Proceedings of the IEEE Conference on Computer Vision and Pattern Recognition, pp. 779–788 (2016)
6. Zhang, J., Huang, M., Jin, X., Li, X.: A real-time Chinese traffic sign detection algorithm based on modified YOLOv2. Algorithms **10**(4), 127 (2017)

7. Lowe, D.G.: Object recognition from local scale-invariant features. In: ICCV, vol. 99, pp. 1150–1157 (1999)
8. Girshick, R., Donahue, J., Darrell, T., Malik, J.: Rich feature hierarchies for accurate object detection and semantic segmentation. In: Proceedings of the IEEE Conference on Computer Vision and Pattern Recognition, pp. 580–587 (2014)
9. Girshick, R.: Fast R-CNN. In: Proceedings of the IEEE International Conference on Computer Vision, pp. 1440–1448 (2015)
10. Ren, S., He, K., Girshick, R., Sun, J.: Faster R-CNN: towards real-time object detection with region proposal networks. In: Advances in Neural Information Processing Systems, pp. 91–99 (2015)
11. Lin, T.-Y., Dollár, P., Girshick, R., He, K., Hariharan, B., Belongie, S.: Feature pyramid networks for object detection. In: Proceedings of the IEEE Conference on Computer Vision and Pattern Recognition, pp. 2117–2125 (2017)
12. Redmon, J., Farhadi, A.: YOLOv3: an incremental improvement. arXiv preprint arXiv:1804.02767 (2018)
13. Lin, T.-Y., Goyal, P., Girshick, R., He, K., Dollár, P.: Focal loss for dense object detection. In: Proceedings of the IEEE International Conference on Computer Vision, pp. 2980–2988 (2017)
14. Qiu, S.: A simple tool for labeling object bounding boxes in images. https://github.com/puzzledqs/BBox-Label-Tool. Accessed 01 Apr 2018

Author Index

Printed in the United States
By Bookmasters